Searching for ORDER IN the COMPLEXITY of Evolving Worlds

## ACKNOWLEDGMENTS

*The SFI Press and the InterPlanetary Project would not exist without the support of William H. Miller and the Miller Omega Program. The SFI Press is also supported by Andrew Feldstein and the Feldstein Program on History, Law, and Regulation, and Alana Levinson-Labrosse.*

# [INTERPLANETARY TRANSMISSIONS]
# [GENESIS]

*Proceedings of
the Santa Fe Institute's
First InterPlanetary Festival*

DAVID C. KRAKAUER
CAITLIN L. MCSHEA
*editors*

*with illustrations by Caitlin L. McShea*

THE SANTA FE INSTITUTE PRESS

1399 Hyde Park Road
Santa Fe, New Mexico 87501

*InterPlanetary Transmissions: Genesis*
*Proceedings of the Santa Fe Institute's*
*First InterPlanetary Festival*
ISBN (HARDCOVER): 978-1-947864-23-8
Library of Congress Control Number: 2019904092


The SFI Press is supported by the
Feldstein Program on History, Regulation, & Law,
the Miller Omega Program, and Alana Levinson-LaBrosse.

Our model of the cosmos must be as inexhaustible as the cosmos. A complexity that includes not only duration but creation, not only being but becoming, not only geometry but ethics. It is not the answer we are after, but only how to ask the question.

—*Ursula K. Le Guin*
The Dispossessed *(1974)*

# Part I: Genesis

# Part II: Proceedings of Santa Fe Institute's First InterPlanetary Festival

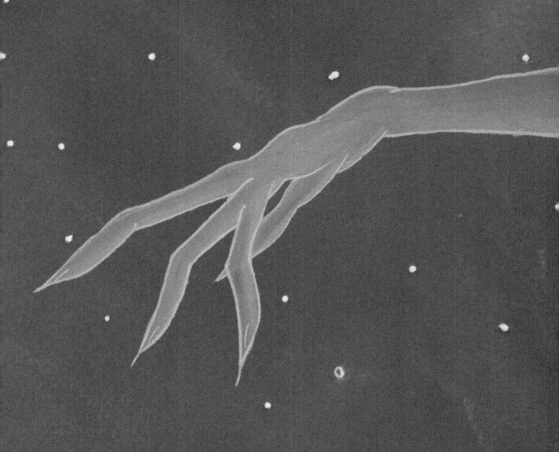

# GENESIS

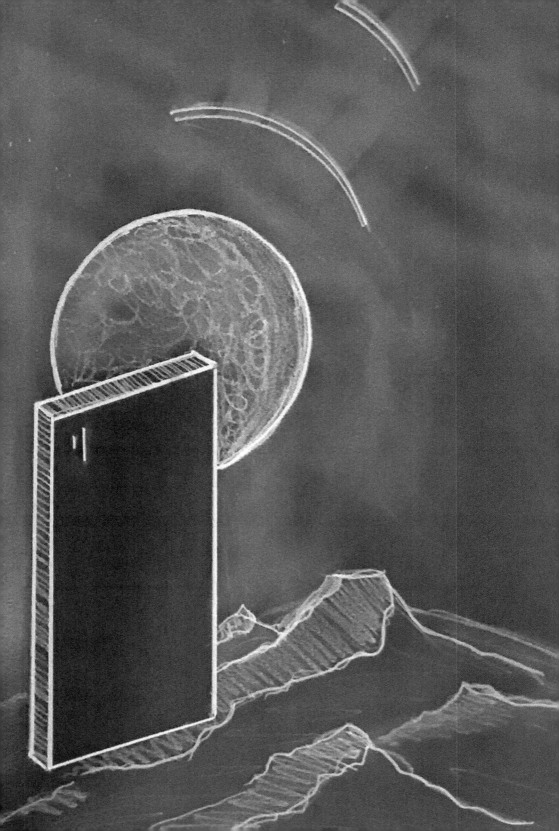

# PREFACE: RESTORING FOCUS AT A PLANETARY SCALE

*Man is an artifact designed for space travel.*

—WILLIAM S. BURROUGHS, "CIVILIAN DEFENSE"

The InterPlanetary Project seeks to engage with pressing problems of today by imagining the challenges of tomorrow. Mark Twain wrote, "You can't depend on your eyes when your imagination is out of focus." InterPlanetary seeks to restore focus at a planetary scale.

What terrestrial challenge could provide greater focus than nourishing and uniting the peoples of the Earth to explore the extrasolar planets? And what great solutions might we find by searching beyond the confines of our Earthly troubles?

To search through outer space we shall need to rise above our inner spaces, the gravest challenges of our time—from reducing disease and economic inequality to managing finite resources and surviving war—and take all necessary steps towards a larger, shared goal: an understanding of life's place in the Universe. Because confronting the challenges of space requires braving and solving the complexities of life.

The InterPlanetary Project is equal parts conference, festival, and research program. The first project of its kind to combine celebration with experimentation and conversation with analysis, InterPlanetary seeks nothing less than a whole-planet project—beyond borders, beyond politics, beyond economics—to activate the collective intelligence of our first planet: Earth.

We began with a conversation at the historic Lensic Theater in downtown Santa Fe, asking our panelists—science fiction authors, scientists, explorers, and artists—to consider, what will it take to become an InterPlanetary civilization? How should we address the most pressing problems of Earth to tackle a challenge at this scale? What will success mean for future generations? What is holding humanity back, and what do we all need to achieve in terms of novel governance, new technologies, shared resources, and global cooperation to achieve this goal?

Santa Fe Institute researchers endeavor to understand and unify the underlying, shared patterns in complex physical, biological, social, cultural, technological, and even possible astrobiological, worlds. Our global research network of scholars spans borders, departments, and disciplines, unifying curious minds steeped in rigorous logical, mathematical, and computational reasoning. As we reveal the unseen mechanisms and processes that shape these evolving worlds, we seek to use this understanding to promote the well-being of humankind and life on earth.

—David C. Krakauer
*President and William H. Miller Professor*
*of Complex Systems, Santa Fe Institute*

So, at the galactic tribunal where you're asked, "What has your species contributed to the universe?" you could stand up and say, unequivocally, "This was an amazing accomplishment and I could tell anyone on this planet or any other one that I thought this was something that was worthwhile."

## Out of this World: Santa Fe Institute Launches
## an InterPlanetary Project with Galactic Ambitions

Science fiction lovers will probably recognize the galactic tribunal trope from any number of graphic novels, books, or films. Santa Fe Institute President David Krakauer introduces the notion of cosmic judges rendering decrees on humanity as part of a thought experiment:

"I'm so interested in the question, 'What can we be proud of?'" Krakauer says during an interview in his office at the Institute where, on this June afternoon, smoke from fires burning in the national forest semi-obscures the vast city and mountain views from Hyde Park Road. Stacks of books crowd most available surfaces—George Johnson's *Fire in the Mind* and *The Book of Trees* by Manuel Lima catch the eye; a model rocket sits atop Krakauer's desk.

"So, at the galactic tribunal, where you're asked, 'What has your species contributed to the Universe?' you could stand up and say, unequivocally, 'This was an amazing accomplishment and I could tell anyone on this planet or any other one that I thought this was something that was worthwhile.'"

Questions such as these, which require considering humanity and life on Earth from an interplanetary perspective, drive SFI's new InterPlanetary Project. It launches July 18 with a panel discussion between scientists, writers, artists, and thinkers whose work all revolves in various ways around humanity's future—in space and otherwise.

InterPlanetary builds on the type of complex, interwoven, boundary-pushing research SFI is known for locally and internationally.

The project also comes at a time during which interest and money for space science is harkening a new golden age for galactic exploration. Among other developments, NASA ushered in a new class of astronauts from a record number of applicants, scientist Stephen Hawking proclaimed humans have only one hundred years left on Earth, and SpaceX technologist Elon Musk published detailed plans for resettlement on Mars.

It's a fitting context, in other words, for InterPlanetary's first panel, for which the starting question is: "What will it take to become an interplanetary civilization?"

That mind-bender encompasses many concerns—from preparing humans for space travel to the hard science and technology required to make the journey, along with aspirational initiatives centered on how we might communicate with other life in the Universe (should it exist). Last but not least: How can humanity tackle the problems of today to prepare for its future?

As a child, Kate Greene always wanted to travel to other planets. She loved both science and stories. "When you're a kid and you think about other planets, it's fertile ground. I loved imagining other planets because the thing I liked the most about science was getting little bits of information here and there and weaving them together into a whole story."

As an adult, Greene ended up on a 2013 NASA-sponsored simulated trip to the Red Planet by luck. She was scrolling through Twitter and spotted a link to an NPR article on why astronauts crave Tabasco sauce. NASA was set to begin studies on astronauts' food habits and, at the end of the article, there was a call to apply to be one of six crew members on the HI-SEAS project—the Hawaii Space Exploration Analog and Simulation, a Mars simulation research project that takes place on Hawaii's Big Island, approximately 8,200 feet above sea level on the Mauna Loa side of the saddle area.

"My palms started to get sweaty, my heart started racing," Greene says. "I had a real physical response to living out this kid dream, so I applied."

Greene applied and was accepted into the program—not by luck. Her background includes work as both a laser physicist and a science journalist. She spent her four months on the project as a crew member and crew writer, contributing dispatches for *Discover* magazine, while also conducting a separate sleep study for NASA.

Day to day, she says, the experience didn't necessarily "feel like living on Mars, and I don't know if living on Mars day to day would feel like living on Mars." A fair amount of her time and, presumably, a fair amount of time for future astronauts who would actually live on Mars, was spent indoors in a habitat that included living, kitchen, work, and exercise space. The project incorporates 20-minute communication delays with the outside world and no real-time conversations to further simulate an off-world experience.

-7-

"For me, it felt more like graduate school, in a way," Greene says. "I studied semi-conductor lasers and spent a lot of time in the basement laser lab. We spent all day trying to figure out experiments. That's what it felt like in a lot of ways: working together, solving problems."

There were times, when the crew would don their spacesuits and walk around in a landscape that is red and rocky and "looked a lot like Mars on the outside," during which Greene could imagine she was exploring another planet. The experience also yielded other rewards, such as contributing research to help future astronauts, and collaboration with new people. Moreover, Greene, who will appear on the SFI panel, says the experience changed her work as a writer, and set her on a new path of integrating her science-writing background with more personal narratives. Her book of essays that came from the experience, *Once Upon a Time I Lived on Mars and Other Stories*, is forthcoming from St. Martin's Press.

Of course, not everyone wants to travel to outer space, or even simulated outer space. But to scientist Lindy Elkins-Tanton, just the prospect of human space travel sparks "intrinsic excitement" that "seems to be deeply rooted in being human."

Director of Arizona State University's School of Earth and Space Exploration, Elkins-Tanton's own research focuses on the processes involved with the formation of terrestrial planets. She has received numerous awards and accolades as both a scientist and a teacher.

In January, NASA announced that Elkins-Tanton's Psyche Mission had been chosen as one of the government space office's Discovery Program projects. Psyche is a metallic asteroid researchers believe is the core of what would have been a planet in the earliest days of the Solar System, and which may provide information about the Earth's own core. The project spacecraft is scheduled to leave Earth in 2022, reach Psyche in 2026 and spend close to two years mapping and studying the asteroid, which orbits the Sun between Mars and Jupiter. "Psyche is probably the only way human kind will visit or measure a core directly," Elkins-Tanton says.

Elkins-Tanton, like Krakauer, staunchly believes that inspiring work—scientifically or otherwise—comes through collaborations across disciplines.

Her own school at ASU is somewhat unusual in science academia in that its faculty, students, and research projects work across a spectrum of interests: from cosmology to earth science to technology. Last year, ASU's president asked Elkins-Tanton to lead an initiative across all the schools for an interplanetary initiative that has, she says, synergy with SFI's project.

"I think what we are doing is really important," she says. "Humans are going to have a space future. Up until now, our space future has been worked on by national organizations like NASA or individual technologists like Elon Musk. We need all the human disciplines to come to bear if we're going to have a human space future. We need the sociologists and the psychologists and the artists and the theater people to really approach what our society is going to be like as we go into space."

As such, Elkins-Tanton says, the initiative ties into what she calls her primary mission. "What we do at ASU that is so complementary to the Santa Fe Institute is we're also trying to create the future of education. We're trying to teach people process and not just content, to prepare them to answer questions we can't anticipate from where we are right now that are going to need to be answered in the future."

One of those questions, as Elkins-Tanton perceives it, is what impact will the discovery of life elsewhere—intelligent or otherwise—have on the human psyche. She references nineteenth-century geologist Charles Lyell, whose work influenced Charles Darwin's human evolutionary theory. Lyell, she notes, "was the first to really clearly explicate and lay out for public consumption the proof that the earth must be very, very old and that there were these reoccurring natural processes that just operated over such a long time scale that it was hard for us to envision what it was."

-9-

This notion, she says, "caused a great crisis of faith. It caused a crisis in meaning. There was a sense that if things weren't laid out by divine providence and we were just subject to this pitiless occurrence of natural phenomena, then we had no meaning." Elkins-Tanton wonders if the discovery of life elsewhere—be it uni- or multicellular—"will give people that feeling that we've lost meaning or the feeling that we've gained meaning again."

Regardless, the prospect that we are not alone in the Universe is fertile ground for scientists, artists, and artist—scientists such as Dario Robleto.

Robleto has spent much of his career thinking on the message people of Earth might want to convey about themselves. A formative moment happened when he was six or seven years old and first encountered the Golden Record, placed on the spaceships Voyager 1 and 2 in 1977 containing images, sounds, and messages meant to capture life on Earth.

Robleto had stayed home sick from school and dialed a 1-800 number NASA had set up for listening to sounds of space as the Voyager made its first approach to Saturn. Robleto called in expecting to hear aliens, and instead heard the Golden Record. "I didn't understand that at all . . . I was so disappointed. It just made no sense to me why NASA would send this into space. I had no idea I was listening to the most beautiful thing I've ever heard."

He's referring to Ann Druyan's brain waves, recorded shortly after she and Carl Sagan, with whom she collaborated on the Golden Record, admitted they were falling in love. Robleto says in many ways his work has been directed by Druyan's actions. "She essentially snuck love on board," he says.

Two years ago, Robleto—who has ended up working with Druyan—became the artistic consultant for the "Breakthrough Message" project, one of Yuri and Julia Milner's Breakthrough Initiatives founded in 2015 "to explore the Universe, seek scientific evidence of life beyond Earth, and encourage public debate from a planetary perspective," according to the project's website. "Breakthrough Message" is a $1 million competition to design a message that comprehensibly captures humanity and life on Earth for another civilization.

Robleto also serves as artist in residence at the SETI Institute. The work at SETI—the Search for Extraterrestrial Intelligence—includes a vast array of research and development projects in the service of finding signs from advanced civilizations in the galaxy.

Based in Houston, Robleto has shown work in both solo and group shows across the country, and has been recognized with a variety of awards, fellowships, and residencies. As an artist, Robleto's practice spans mediums: He works in sculpture, paper, print, and more. As an artist with both a background in and a passion for science (he was a biology major before switching to the arts), he prefers the term "transdisciplinary" because it fully encompasses the degree to which his projects are intertwined with rather than merely referential to scientific issues and phenomena.

"When I made the switch from science to becoming an artist, it never occurred to me to not bring that background with me," Robleto says. "I find the tension between them fascinating . . . The common seed is that both originate from the quest to increase the sensitivity of our observations."

Complexity science, which guides SFI's work, particularly interests Robleto. "I'm really drawn to the issue of scale in complexity science, how scaling up and down reveals a different set of information that wasn't apparent from the previous scale. That's not unlike what artists do."

The question of scale connects with another driving concern Robleto brings to his work and that has been a research focus at   –11– SFI: altruism.

"Competition and destruction," Robleto says, "dominates so much of all our topics." His personal quest "is poetic and scientific proof we can be cooperative at some fundamental level," which makes as much evolutionary sense as the belief that competition drives human existence.

After all, over the span of time, "the only civilizations that will survive are ones who figure out how not to destroy ourselves. If we ever found another civilization, just the fact that they're there, that means altruism won out."

Return to Krakauer's original thought experiment: What has humanity done for which it can be proud? Krakauer believes that while specific citations would vary from person to person, most would likely name artistic or scientific achievements rather than, say, "human decency," although he believes the latter also would be legitimate.

The question ties into Krakauer's own background as a mathematic biologist, and one of the many thoughts that prompted the InterPlanetary Project as a whole. We are living in a time in history, he argues, filled with "incredibly positive things," among them the ability to connect with other people all over the world, progress in our understanding of the environment, "the second great space race," and many other accomplishments. These signs of progress are occurring alongside other signs of the worst time in history, what Krakauer characterizes as a "stupidity pandemic, and the deepest political distrust and the worst kinds of prejudice." Interdisciplinary work, such as the InterPlanetary Project, provides an opportunity to consider how and why such darkness and light can occur together—and perhaps yield some solutions.

"I'm very sensitive to this idea that an interplanetary concept would be considered irresponsible and that we're running away from the problems we're facing now," Krakauer says. "It's the opposite. Friends have asked me, 'Don't we have real problems with income inequality and spoiling the environment?'" Yes, Krakauer says. And "if we really want to settle a colony on Mars, we sure as hell better understand how to create a social system that doesn't fragment within a year, we better understand planetary cycles and we better understand how we're going to grow food in an inhospitable environment."

Re-contextualizing some of these challenges within enlivened discussions is a way of "injecting playfulness . . . and hedonism" into serious topics.

Changing the lens through which we view our civilization and its workings might be one way (if an overly simplistic one) of considering panelist, former SFI president, and distinguished professor Geoffrey West's work. For much of his career, West, a theoretical physicist, focused on fundamental questions of physics: the cosmological implications of the evolution of the Universe, the nature of dark matter.

For the last 15 years, his research associated with the Santa Fe Institute has been looking, from a physicist's perspective, which he describes as "quantitative, analytic, computable, and therefore predictable," at "the phenomena that go on on this planet. In particular, the questions to do with the generic principles giving rise to life and laws of life, the regularity that underlies the extraordinary complexity of life around us, social life, the socioeconomic life, the life we've created on this planet." (West's most recent book, released in May, is titled *Scale: The Universal Laws of Growth, Innovation, Sustainability, and the Pace of Life in Organisms, Cities, Economies, and Companies*).

Just as Krakauer and Elkins-Tanton both emphasize the need for work across disciplines, West points to the way in which an interplanetary viewpoint can create a more holistic picture "in which everything is recognizing that everything is interacting with everything else and everything is dependent on everything else."

If "changing the world one planet at a time," as Krakauer describes the project's aims, sounds overly ambitious—that, too, is part of the point. "Cynicism is boring," Krakauer says. "Gloom has its own aesthetic appeal . . . and optimism can be a little boring at times, but it is an optimistic project in that it says ingenuity is unbounded and understanding in many ways is unbounded and there's so much more to know."

—*Julia Goldberg*
*adapted from* the Santa Fe Reporter, *July 11, 2017*

# INTRODUCTION:
# THE RECURRING LA CHOZA CHAT

"So, what do you do here in Santa Fe?" the stranger next to me asks, excited to converse with a local over his green chile stew.

"I work at the Santa Fe Institute."

"Oh, I've heard of that. It's that big *[insert "lab" or "think tank" here]*, right?"

"It's a complexity science research center and educational facility."

"Uh-huh, uh-huh. And what do you do there?"

"I direct the InterPlanetary Festival!"

"The what, now?!"

"The InterPlanetary Festival! It's . . ."

I won't take the time here to give you the bar-top spiel on the InterPlanetary Project and its mission, because David already has. But it's worth noting that this conversation is usually met with skepticism from my interlocutor, so smash-cut to the reliable old gibe:

"Why are the eggheads up there wasting their time with a space party when they should be focusing on the problems here on Earth?"

"We're throwing a space party *because* of all of the problems on earth," I silently scream to myself as I sip my margarita, and search for some civil composure within.

◇

No, I'm only (half) kidding. I completely understand that, upon first blush, a festival seems a frivolous pursuit by any organization, but most especially by a nonprofit organization primarily focused on scientific research. What on Earth could a party possibly accomplish? I'm totally enamored with that question because, when I try to answer it, I can't help but recall some of the innovative dance moves I've witnessed, the games invented, music written, snacks concocted out of desperation, and the out-there ideas explored by friends of mine in the uninhibited space of a celebration.

Humanity is on the brink of multiplanetary habitation. We've developed robust systems for communication and transportation, among other things, and, in my lifetime, ours will be the first species (that we know of) to occupy more than one rock. When tasked to think about how best to approach this inevitability—how to build a thriving civilization elsewhere on austere and unforgiving terrain, and how to repeat what we've done, or surpass what we've done, that one time we did it on Earth—where better to engage in such a thought experiment than the audacious and mettlesome occasion of revelry? What *can't* a party accomplish?

The Santa Fe Institute has been working on the outermost edges of science since its inception nearly thirty-five years ago, and the researchers there aren't afraid of very big or very hard problems. They work across (some might say "contra") disciplines to gain perspective on questions you might not expect science to tackle. "Why don't we live forever?" "Why does time only move forward?" Ask the biologist, the physicist, the investor, the poet. SFI is really good at putting motivated individuals of disparate backgrounds together to contribute their insight on these problems. After all, one of the most incredible aspects of humanity's collective potency is the fact that, individually, we're all unique. The Institute continues to succeed in its quest to seek order in the chaos of evolving worlds because of this approach, because of the peculiar and productive properties of a gathering. So, naturally, in an effort to change the world one planet at a time—InterPlanetary's ultra-modest mission—SFI decided to host its largest gathering yet. Not a meeting or a workshop. Not a conference. A multi-day outdoor festival, free for all, and streamed globally.

The festival as a medium allows SFI's perspective on planetary success—the complexity of our living planet and the systems that govern its fitness—to be contextualized, considered, and criticized at times by groups of very cool and very different people, audience included. It's a truly multimedia experience too, because scientific researchers aren't the only ones exploring these themes. Game developers, filmmakers, musicians, policy-makers, advocates, authors, and artists from all over are creating wide-ranging opportunities to confront these pressing issues of tomorrow. Will AI prove to be a great benefit to us in the future, or will it bring about our ultimate demise? Participate in a conversation on various forms of intelligent systems; witness an algorithm attempt to paint a portrait; watch a terrifying, though awesome, dystopian action movie; read an acclaimed sci-fi novel on the subject: choose your own adventure and decide for yourself.

A festival is not the place for experts at a podium to spoon-feed you an opinion. A festival is a place to relax, enjoy, and indulge. You come to a festival to dance to live music performed outdoors on a stage, through the air, under the warm sun. You sway in the comfort of a thin t-shirt, you hang out under the shade of a tree with your dog, and you drink an ice-cold beer. But what sets the InterPlanetary Festival apart from others is that, while you're in the midst of your pint or your ice cream cone, you're met with accounts from an architect building lunar habitats smaller than a porta-potty or a virtual reality artist who mimics the experience of falling between Saturn and its rings. You close your eyes and transport yourself to a festival in space in the not-so-distant future. You try to embody that thick, stiff spacesuit; you try to navigate the lacuna. You attempt to stomach the toothpaste-textured meals washed down with powdered orange drink. Your muscles begin to atrophy, and you lose every untethered belonging. You tweet no one. You hear nothing. You have to monitor your oxygen closely because you've only a limited supply.

A microphone's feedback squeals on stage, and you snap out of it. You feel that desert breeze on your face, and remember that you're grounded (for now!) in sunny Santa Fe, New Mexico, United States, Earth.

It's easy to take for granted all that we enjoy as human Earthlings. The atmosphere and ecology of this planet furnish so much, and we've been clever enough to create myriad creature comforts on top of that. For better or for worse, we are a remarkable species, but our evolution into space-faring scouts is the result of an intimate link with Earth and all of its distinct conditions. We are, at this very moment, capable of traveling through the interplanetary ether, either on a touristic moon cruise or on a construction contract to build a lunar refueling station for those on an even longer trip to Mars. It's a testament to the incomparability of human ingenuity and to those brave souls who have devoted themselves to penetrating the great and often frightening unknown. We're able to look up to the sky, which is no longer the limit, and contemplate an off-planet future. It isn't an unreasonable or irresponsible investigation at all. And it shouldn't suggest that an off-planet future is the only future we have to look forward to. But its actuality will be an unprecedented accomplishment all our own. Isn't that worth celebrating?

In order to produce a vibrant, dynamic future for ourselves elsewhere, and in order to maintain (or optimize) such an existence here, we all will need to grasp everything that underlies and influences a living planet—and Earth is the only one we have. We should do everything in our power to promote its longevity, if for no other reason than the fact that we've become really good at producing beer, and Mars malt is a long way off. Intuiting and appreciating one's agency in the Universe is a task for every single citizen of this planet. It doesn't have to be hard, and it shouldn't be boring. In fact, SFI demands that it be fun! So the eggheads up on the hill have decided to throw a space party.

-19-

"Oh, wow, that sounds really interesting."

"It will be. You should come!"

*—Caitlin L. McShea*
*InterPlanetary Festival Director and*
*Manager, Miller Omega Programs,*
*Santa Fe Institute*

# ROUNDTABLE:
# WHAT WILL IT TAKE TO BECOME
# AN INTERPLANETARY CIVILIZATION?

*A gathering of intergalactic luminaries took place in Santa Fe, New Mexico in July 2017 to discuss and shape the future of the Santa Fe Institute's InterPlanetary Project.*

**DAVID KRAKAUER** Let's get started. First of all, Cormac [McCarthy] is on his way. We said, "He knows. He knows." I said, "Let me just give him a call." "Hello?" "Hello. It's tomorrow, isn't it?" So he's en route. He'll be here in his very large truck, so expect casualties.

Okay, I want to get a little background on why we're here and you'll see it's fairly free-form. This should be free-form. The primary purpose of this little group here is to get to know each other. So tangible outcomes of the sorts you might be accustomed to in meetings are not particularly interesting at this table. That's number one. Everyone should have an opportunity to say something. If someone is saying too much, I will shut them up and give everyone a chance to speak.

I want to discuss deep ideas, so it's going to be an intellectual conversation at this table to get us all sort of limbered up for this evening. Allen looked and he said, "You're all nervous." I said—

**ALLEN EZ** No, I was just wondering how Lou's going to fare.

**D. KRAKAUER** Yeah, I think Allen is going to ask him about quantum mechanics at some point.

**A. EZ** Only if you want to see a grown man cry.

**D. KRAKAUER** Towards the end we'll talk a little bit about logistics and implementation-type stuff. Again, that's not so interesting here, and then we'll get into October and June and where this whole project is going. A few things first: What is the InterPlanetary Project about? In a nutshell, it's understanding living planets. Is it the evolution of living planets? The laws on living planets? Physical law? Biological law? Social law? Economic law? It's understanding living planets and their long-term prospects, including their collapse. It's not just planets out there; it's our own, and that's very important. It's the only one that we know of that's living, but InterPlanetary gives us a rather unique perspective on that.

Don't feel the need at any point to map this onto any kind of project that you're familiar with. If you do that, that would be a negative. We're not interested. Everyone has experience of this sort of thing, but the last thing we want to do is start by saying, "Oh, this should be more like X." It shouldn't be like anything. It should be something creative that we conceive of, and of course we're going to draw on our experience, but I'm just not interested in turning it into something that already exists.

There are two slogans, if you like, for InterPlanetary; the one we use is "Changing the world one planet at a time." It's very modest! The other one I just realized last night could've been "The sky is not the limit," which would have been quite natural. It's a good T-shirt slogan, but it doesn't quite capture the special element that the previous slogan does.

The final thing I do want to say is: What is it? I have some metaphors, and we'll open this up into conversation in a second. The first thing it seems to me is that it's a lens. If there were a metaphor for what the InterPlanetary Project is, it's a lens. Sandra, you'd like this: Think telescopes. What do they do? They make invisible things visible, and they bring things into focus. Lots of issues, I think, are brought into focus by thinking about them from an interplanetary perspective.

Slightly more nerdy—I'm sitting next to Neal, so I can get away with this—it's a simulation engine. Think of it as a simulation engine. If you think about game engines, you create artificial societies, with artificial rules, with emergent behaviors. The InterPlanetary Project should be something like that, but in real life.

Finally, it is a participatory project, meaning that it produces things. It's not like Burning Man or Comic-Con, where you go to enjoy yourself exclusively. Out of it come films. Out of it come books. Out of it come movements. We should think about it as an engine for generating real products of various kinds, not necessarily commercial, but also intellectual, and so forth.

That's sort of it: lens, simulation engine, project. Which means that it continues to exist in between events, and so, again, don't just think of this as an event. Think about what it would be online. Think what it would be away from Santa Fe, God knows, anywhere else in the world. So that's the background.

What is it? It's about living planets, and their ambitious thinking. I know that Neal was involved at ASU in Project Hieroglyph. That's sort of interesting because it did overlap, I think a little bit, with us, in the sense that it was asking very big questions, daring to ask very big questions. What makes this different is that it's fueled by complexity science, fueled by the Santa Fe Institute, which means that at the base are really deep ideas, and there are deep ideas that I think people would be very interested in learning about, that could make a difference in their lives.

Deep ideas don't have to be dull. Deep ideas have to be transformative and fascinating and fun. It's part of the mission of InterPlanetary to do that.

With that said, I'm going to shut up and have you introduce yourselves around the table briefly, and then we should just jump into some deep questions. If you hesitate, then I'll throw them onto the table because there are some things I certainly do want to discuss. You know who I am.

**NEAL STEPHENSON** I'm Neal Stephenson, author of science and historical fiction books from a kind of science technical background, and I suppose my best credential for being here is that I was involved very early in Blue Origin, which is the private space company that was founded in 1999 by Jeff Bezos. I haven't been directly connected with them in about eleven years, so I don't in any way represent them, or speak for them, but that's kind of the context of my being here.

**D. KRAKAUER** And Neal is a Miller Scholar of the Santa Fe Institute.

**N. STEPHENSON** Yeah.

**D. KRAKAUER** Incidentally—I should say this because I was remiss not to have mentioned it—this is all happening in large part because that gentleman over there, Bill Miller, founded a program at the Santa Fe Institute called the Miller Omega Program that asks after the most ambitious things that the Santa Fe Institute could possibly do for the planet. And so without Bill, there would be nothing, including Neal, so thank you very much.

**BILL MILLER** Neal would still exist without me.

**D. KRAKAUER** Yeah, you'd exist, but you wouldn't be here.

**N. STEPHENSON** I'd exist in Seattle.

**SANDRA MOORE-FABER** Sandy Faber, Professor Emerita of Astronomy from UC Santa Cruz and a staff member of University of California Observatories, builder of telescopes and telescope instrumentation. Why am I here? I'm here because thinking about astronomical knowledge and its value has made me wonder if it could be of profound value to everybody else on the planet in helping us think about the future of the planet. Now, we know from astronomy basically how we got here and we can predict pretty much what the next hundred million years is going to look like on Earth, and the message is that it looks good.

We're facing some really difficult challenges right now, trying to live within our means on our planet. I'm wondering if communicating this beautiful cosmic future might be used to refocus people's attention on the possibilities for our species, if we got our act together.

**PAUL BUCCIERI** Hi, my name's Paul Buccieri. I oversee A&E Studios on the television side. I also oversee A&E Independent Films, and I am the president of the portfolio group that encompasses A&E, Lifetime, History, Element, and FYI. I'm just here to learn. I'm also here to introduce some commercial possibilities of anything that's interesting, but mostly I'm here to listen and learn.

-25-

................................................................

We're facing some really difficult challenges right now, trying to live within our means on our planet.

................................................................

**MISTY TOSH** My name is Misty Tosh, and I've been working with these guys for a bit, putting together just good stuff on tape, documenting everything that they're doing, and I suppose the reason that I'm here is to make everything everyone in this room is doing very entertaining if I can, and to get it all down for posterity.

**D.A. WALLACH** I'm D.A. Wallach. I'm a recording artist and a technology investor. Within my portfolio, the only thing that is directly relevant to InterPlanetary life is the company SpaceX, which is obviously explicitly trying to colonize other planets, and I'm a huge fan of the Santa Fe Institute, so I'm thrilled to be here.

**CYNDI CONN** I'm Cyndi Conn, executive director of Creative Santa Fe, and I'm here to support the Santa Fe Institute in how this project moves forward in the future on the local side.

**ADAM KOHLER** Hey, I'm Adam Kohler, I'm a health-care and life-science technology investor and I cofounded the Parker Institute, and I spend a lot of time thinking about the incentives that help large scientific projects come to life. I'm here to listen, to learn, and to see if there's any way I can help.

**THOMAS ASHCRAFT** Thomas Ashcraft. I have an observatory about twenty miles outside of town. I'm a radio telescope maker. I observe the Sun and Jupiter, transient luminous events, and the mesosphere.

**JESSICA FLACK** I'm Jessica Flack. I'm a professor with the Santa Fe Institute. I work on collective computation, so how nature and society solve problems through information processing, which seems like it might be required for the InterPlanetary Project.

**B. MILLER** I'm Bill Miller. I'm an investor and chairman emeritus of the Santa Fe Institute's Board of Trustees.

**LOU WALLACH** My name's Lou Wallach. I'm a television producer with National Geographic Television. Like Paul, I'm here to mostly listen, with the idea of how to make all the things that we're talking about here, the questions and some of the ideas that come about, how to make them accessible to get the message out to a worldwide audience.

**CHRIS KEMPES** I'm Chris Kempes. I'm an Omidyar Fellow at the Santa Fe Institute. I'm trained as a physical biologist. I'm interested in how life forms, what happens to life once it does form, and how life interacts with planetary environments.

**ANDREA MEDITCH** I'm Andrea Meditch. I was trained as a linguistic anthropologist, but what I really wanted to know was, how do we tell stories to each other and particularly nonfiction stories? Like a lot of you guys, I was actually at Discovery for many years, helped build the first film unit and also helped launch the first content website in '94. You would have me to thank for all the Crittercams, except when you have dial-up, you really can't watch a panda move, but we thought about how to engage.

For me, it's how do we gain perspective, whether in the past or in the present or the future or from space or on our own planet? How do we develop empathy? How do we engage people with our stories? I'll stop there.

**A. EZ** I'm Allen Ez. I'm head of element for National Geographic on the television side. We are the largest entity on the planet for the promulgation of public knowledge around science and scientific education. We have a monthly reach of about 750 million people. In the digital space, we are the largest brand on the planet, so basically the power rankings go the Kardashians, then National Geographic, when you're on social. That tells you a little something about human- -27-
ity's priorities and maybe why we should be sitting here today.

**BRENDAN TRACEY** I'm Brendan Tracey. I'm one of the postdocs at the Santa Fe Institute. My background is in aerospace engineering, but I've been the weirdo in that sense, and I try to work on bringing machine learning tech into engineering design. In addition to the big questions that we'll hopefully talk about today, one of the things that excites me about this is this big collaboration, so in addition to my academic work, I've also been in the open source coding domain, and one of the cool things is, there's this project that I'm involved with. When we started, I was living in California. The second person lived in Adelaide, Australia. The third person lived in Japan, now lives in Germany, and the fourth person lives in France. None of us have ever met each other in person and yet, because of this collaborative nature of the internet, we're able to construct something that I'm really proud of.

One thing that excites me is how we can take these kinds of opportunities that are enabled by the internet but then also try to bring them into a common location to see what happens and what you can construct.

**DARIO ROBLETO** My name's Dario Robleto. I'm a visual artist from Houston, Texas, and I became an artist mainly because of the complete misunderstanding and misreading of the Golden Record when I was a little boy. I thought we had made contact. I thought we could hear that, and I was very disappointed in what I did hear, but there was one recording onboard that has in many ways set my path as an artist. And I'm still working on it to this day, which is to tell the story behind the heartbeat that's onboard. I have for many years been trying to tell the story of our quest to record our hearts. Of course, that means something emotionally, but also scientifically. For example, where is the first heartbeat that was ever recorded? How did it happen and why? It's a beautiful story that I'm trying to piece together that all culminates with the launch of the Voyager and the Golden Record.

-28-

I'm an artist in residence at the SETI Institute and artistic consultant to the Breakthrough Initiatives. I also have a great interest in the problem of message design and communication, which is maybe why I'm here.

**GEOFFREY WEST** My name's Geoffrey West. I'm on the faculty at the Santa Fe Institute. My training and most of my career has been spent doing I suppose what you could call fundamental physics—quarks, neurons, and string theory and dark matter, and so on and so forth. But at some stage when I made a transition to the Santa Fe Institute, I became very interested in similar fundamental issues in biology. Most of the work I do is very much big picture. I'm asking what I believe are big questions in biology and now questions of social organization and understanding.

It sounds somewhat pretentious, sort of "how nature works," but how do *we* work? This collective. So I think the most fundamental question is, how do we understand cities and organizations and their potential threat to the sustainability of the planet—especially when that is the origin of all the great things we do on this planet, this urbanization? And yet urbanization is also the driver of all the tsunami problems that we have to face from climate change, to questions of risk in financial markets, to energy availability, population growth, and so on. All of these are bound up with urbanization.

And so maybe my biggest passion—I have two big passions, really. One is, I have a morbid interest in aging and mortality, but at all scales. That means not just at my personal scale, but the scale of human beings and animals. Why we age and why we die. To understanding the underlying principles, as a physicist, to be quantitative and predictive. Also, the question about the aging mortality of Google and Microsoft, or of San Francisco or Beijing, and ultimately the question of the aging and mortality of the socioeconomic enterprise which we've been engaged in for this little blip in time, and whether this is just some fantastic experiment of natural selection that lasted ten thousand years, or whether it's going to lead to some glorious future where we do move into interplanetary space as well as into *intra*planetary space. That's what I have, in my aging years, dedicated myself to.

**KATE GREENE** I'm Kate Greene and I was trained as a laser physicist. You were talking about lenses; that really resonates with me. I remember spending many months calibrating the lens position and choosing the right one and making this amazing optical train, and the challenges that that entailed. So it seems like what you're saying, David, is something that I've had some experience with. But, right now I'm a writer. I went from laser physicist to writer, specifically, I'm a science journalist. I'm probably here because I was on the HI-SEAS Simulation, which was a surface Mars simulation, where for four months I, along with five others, lived as if we were astronauts living on Mars. We couldn't go outside unless we were wearing simulated spacesuits, and we couldn't talk to people back home without a twenty-minute delay. We really understood the isolation aspect, in that regard.

Something about that experiment that really stuck with me was how quotidian a Mars mission would actually be when it comes down to it. We're asking ourselves really big questions, but when it comes down to it, there will be people experiencing that journey and people's lives are made of small moments, ideas, incidents, emotions. I'm really curious about how the mundane everyday plays into the big questions as well. I think that there's a conversation in between those scales. I'm interested in exploring that.

**SCOTT ROSS** My name is Scott Ross and I was involved in computer imagery way back when. I was the head of Industrial Light & Magic when we transitioned from analog to digital. I ran Lucasfilm and Skywalker and all of the other divisions of George's company. I left and formed a company called Digital Domain with James Cameron and Stan Winston and ran it for about thirteen years, and won about nine Academy Awards over that period of time. Why I'm here is because Krakauer called me on the phone. I couldn't quite figure out why, and now, having listened to what you all have said, I have a feeling I understand why I'm here, and that's to learn from you all, so thank you very much.

**D. KRAKAUER** All right, well, what a good group. What a modest group. I don't believe it for a moment! Okay, so actually, let me start because last night Chris interviewed Neal at the Jean Cocteau Cinema, which is owned, by the way, by George R.R. Martin, who is a local resident. And, you made a really interesting point, Neal. I think it would be a great place to start, if you would be willing to reprise that thought, the very funny narrative on Michigan versus Mars.

**N. STEPHENSON** Well, I've been reading *Cadillac Desert*, which is a book that's been around for a few years, but it's a great account of the fundamentally unrealistic efforts we've made in this country to get a civilization built in the arid Southwest, where there simply is not enough water to support such a civilization. I recommend it because it has all sorts of resonances I wasn't expecting with our current political situation. I guess the bottom line is that it's not working. We've made these Herculean efforts to divert water to Los Angeles from the Boulder Dam, or the Hoover Dam, and that wasn't enough. The events that are the subject of the movie *Chinatown*, the kind of annihilation of the ecosystem of the Owens Valley in order to get water for LA, figures into this, and there were a whole bunch of utterly demented plans to divert water from the Columbia River, from Alaska, from the Great Lakes, just to get more water into Southern California.

The result is that a lot of these efforts have backfired. A lot of places that have been successfully irrigated are now in ecological trouble because irrigation brings salt to the surface and turns the soil into an alkaline, salty substance that won't support plant life. For me, it's interesting to read all of this in the context of conversations about terraforming other planets, because if we can't really get it to work very well in places that can easily be reached in a car or a plane, places with air to breathe and at least some water in the soil, and so on, then I don't quite understand why we think that we can make it work on Mars.

-31-

I ended up—not to be totally negative—I ended up getting into the topic of Mormons, which is a thing that's mentioned in *Cadillac Desert*. They came to this part of the world partly to get away from lynch mobs that were chasing them out of everywhere they went, but partly for a positive reason of having a sense of mission, of wanting to colonize and build a civilization in a new place. They got very good at irrigation. They were early terraformers in a way. This is just not a great big coherent conclusion so much as a set of reflections that I thought were sort of interesting and relevant to the topic of terraforming. And what I'm kind of saying is, I personally doubt that it works, that terraforming other planets works or makes sense unless people go there with a sense of mission in the same way that Mormons and other religious groups have sometimes gone out into godforsaken parts of this planet to build new civilizations. Does that cover it?

**D. KRAKAUER** It does, and so, to throw it out to the table, what I thought was really captivating about it was those two remarks. One was that there are challenges on our own planet that are apparently much more manageable, but that we have failed in, yet we have this extraordinary optimism for what we might be able to accomplish in much less hospitable environments. And somehow, there's a desire to do it and that desire should not be fundamentally economic or practical, but actually ideological. Ideology in the end wins. It's the stronger motivating force.

That's one of those beautiful lenses! By focusing on terraforming another planet, you have to think very carefully about the extraordinary failures of terraforming our own planet. Does anyone want to comment on these issues or raise this discussion? I think it's one of the primary criticisms, by the way, of a project like this—that it feels irresponsible, sort of abdicating your responsibility to your own Earth.

**A. MEDITCH** I have a comment and a question, I guess, for everybody, because one of the other things Neal said last night was that it's also driven by obsessive dictators and by war. That mission becomes driven by destruction, if you will, as well as by mission. Because my background is in trying to use nonfiction stories to get people to have perspective on other people's paths, I guess the fundamental question for me is, are we looking outward only, or are we using the perspective from out here to look back? For me, that's the richest venue, but there are a lot of people here on the other side of the house who are looking out to other planets. I throw that out as a question for the table. It's something that I feel I need to get my own head around in thinking about how this group goes forward in conversation. Obviously, they're related. Is it looking out, or is it also being out in order to be able to look in and gain new perspective?

**K. GREENE** I attended an interstellar workshop headed by Mae Jemison, former NASA astronaut, and she has some crazy, out-there ideas. She brings together a lot of people who like to think about a hundred years from now, and whether we can get to another star system. She deals directly with this criticism. You know: "Why in the world would we want to consider interstellar travel when we're just really stuck in lower orbit and haven't even put ourselves as a presence on Mars?" And her response to that is, by overshooting, potentially you come up with a ton of different creative solutions that will always be applicable to your situation here. I found that really heartening.

To your point about looking, when you go out, to look back, I think that's exactly right. When you start broadening your field of view, that of course encompasses yourself. You find out what home is when you've left home. So I think overshooting to terraform Mars could solve some problems here on Earth.

**D. ROBLETO** Do you feel in your remarks, the Mormon example, that the proper ideological framework is not in place for terraforming on Mars? That we don't have the zeal, currently?

**N. STEPHENSON** I think there are some people who do have it, -33- some nerdy people who are fascinated by the idea of just going out and doing it for its own sake, but it's going to have to rely on that, not on any economically sane argument.

..............................................................

## By focusing on terraforming another planet, you have to think very carefully about the extraordinary failures of terraforming our own planet.

..............................................................

**A. EZ** Hasn't that always driven the desire for exploration? If you look at early man . . .

**N. STEPHENSON** Yeah.

**A. EZ** Okay, so you probably know where I'm going here, but if you look at early man—if you look at the distribution of islands and continents—there was a point at which somebody said, "We're going to build a boat and we're going to travel from wherever to Australia!" And no sane person would have done that because you were probably talking about two or three hundred miles at a time when there was no navigation. There was no sense of being able to map the world around you, but there appears to have been a sort of madness gene that pushed people at select points in history to take those leaps of faith. Had they not taken those leaps of faith, then we'd still be rubbing around in the dirt with sticks.

I think maybe what we're trying to do is make something clean and rational which hasn't always been clean and rational but has led to incredible benefits. To your point of overshooting, you have to live in the poetry of the infinite in order to get to the pragmatic nuts and bolts of what you need.

**S. ROSS** So I guess the impetus, earlier to Neal's point, was it the early Christians' desire to go to Skellig Michael, to climb Skellig Michael on an uninhabitable piece of rock in the Irish Sea, because they had that desire to do it, or was it because they had a religious purpose to do it? I would contend they had a religious purpose. They thought they were getting closer to God. Or was it the Mormons' need to move to Utah to escape what people were doing to them around the country? So there might be two reasons for that movement. One is pure discovery and madness. The other is just desperation and need. The question becomes, which one is it?

-34-

> There appears to have been a sort of madness gene that pushed people at select points in history to take those leaps of faith. Had they not taken those leaps of faith, then we'd still be rubbing around in the dirt with sticks.

**A. EZ** Also, at what point do we as a species collectively decide we could reach that point of desperation? Because there is a bifurcation at the moment in culture between people who see an existential threat to humanity because of what we're doing to the Earth, and those who are quite happy to just carry on living a life that feels more like complacency.

**S. ROSS** I guess my naïveté was that I was never thinking, as you brought up the concept of the project, that it was necessarily about terraforming Mars. That was not my thought. My thought was, at some point, in some way, that through interstellar travel we'd be able to discover environments that were like the Earth. So the effort to be able to migrate to those planets would be considerably less than an uninhabitable environment like Mars is to the human species.

**D. ROBLETO** As far as starting the conversation, I'm curious as to why you want to start here? Is it as a cautionary point to make within the analogy? Do you feel people working in these fields don't already have that cautionary point in mind?

**S. MOORE-FABER** I think this is a very good place to start the conversation because it's about motivation, and as we went around the table, almost nobody mentioned a long-term goal. I think the essence of our work here is to think about the long-term future of the human race as it might be realized through an interplanetary enterprise, but I think that's a very, very deep question. And as Allen just said, there are people who somehow saw some long-term reason to restrain resources versus other people who saw no reason whatsoever to restrict their consumption. That's all about motivation.

As I see it at the moment, the long-term problem is, can we as a species develop a goal that will motivate us, analogous to a religious goal—maybe without religion, maybe it's spiritual, maybe it's a sense of purpose, I don't know. But I tend to agree with Neal that without something like that, something that's ennobling and deeply motivating, it would be very difficult to restrain the worst angels of our nature today. We need a reason to support the better angels of our nature today.

**A. MEDITCH** I was really struck by the combination of what you're saying and what you said, Kate, about seeing what home looks like from another perspective. I think, again, I'm on the side of the coin that says the reason to go out is to save one planet at a time, and we need to start with this one. And that being beyond the system means that you can have a better view of it and that you can see yourself as a greater unified whole, and you may be able to inspire people to take action because they can see themselves as part of a greater system. That seems to be something we just can't get across: that we are all part of this one thing. This sense of home to me is a key piece.

**D. KRAKAUER** You know, Dario, your question is interesting. I thought what was more interesting about the Mormons was their intensity and commitment to principle necessary to do anything great. Whether it's an abhorrent exploratory gene, or what have you. That always strikes me as very interesting. I feel that most institutes or places I know that are really intriguing have a very strong sense of mission, or whatever you might want to call it.

There's another dimension I'll raise. Again, I think we should speak about anything we like, but I want to push this a little bit. It has to do with the kind of science that we do at SFI. We don't do disciplinary science. As far as we're concerned—not as far as we're concerned, as far as factual history is concerned—the disciplines were invented in Prussia and Hanover some time towards the end of the eighteenth century. That made sense then. It doesn't make sense now.

That doesn't mean there isn't extraordinary depth in disciplines. There is. But there are problems with the disciplines and those structures of knowledge that we've inherited. They're an impediment. And SFI was in some sense created to transcend the boundaries of the disciplines. So people always ask, "What is complexity?" Well, I visited a high school and the teacher asked me to talk about interdisciplinary science, which to me is the biggest yawn in the Universe. So I was in this position, but I hate the idea of "interdisciplinary." I hate the idea of "multidisciplinary," too, because they genuflect towards the altar of the discipline, and I don't want to use these words.

So instead I asked the kids, "Who would you put on a mission to Mars?" And they said, "A doctor." "A geologist." "An astronaut." "A politician." Whatever it was. They recognized, instantaneously, that the particular problem defined the sort of diversity of expertise and insight that was required. That's how we should be thinking about the work. I think what happens is all of these problems that we at this table all care about get channeled into the structures of knowledge that like to dissect them up and atomize them in such a way that it becomes, "Oh, I'm a cosmologist. This is a cosmology problem." And the shows you [produce], which are actually beautiful, illustrate precisely this principle. So *Origins* or *Cosmos*—those are exactly what I'm not interested in.

They're great, but they do exactly that. Where's the space for the discussion and the dialogue that does not respect those disciplines? That's something I want to push out to the table as a discussion. How do we articulate that vision of the future that isn't given to us by universities and academics? What's this new way of thinking about the world?

**A. EZ** It doesn't exist within fiction. Hasn't that always been the pattern? Neal can probably speak to this far greater than I can. But science fiction writers have imagined worlds, and in doing so they've created a new language and then real life gets caught up in it. Because, to your point, it's a syntactical or a linguistic problem. If you've only got these terms that you can use in order to answer or ask questions, you're limiting the way you're asking those questions. Therefore, you need people to provide you a new vocabulary, and storytelling traditionally has done that.

**D. ROBLETO** I'll start with the obvious thing you all expect me to say, which is the role arts can play. To your point, I feel I largely became an artist because I realized I didn't want to pick one of my passions—I wanted all of them. Why did I have to pick? So art, to me, is this umbrella term to just put all the passions under, and then just let it all break out, however it wants to go. And it's telling to me that bringing an artist on the mission didn't come up, because I would always argue that an artist excels at the very chaotic bumping of knowledge that you just described, and that artistry is in the unexpected connections across fields.

I'd be curious what you think about "transdisciplinarity."

**D. KRAKAUER** I have the same feelings about it.

LAUGHTER

**D. ROBLETO** Okay. Well, let me make a case for why I choose to use that word in my own description. One of my biggest pet peeves about art-science collaborations is in the spirit of what you just said, which is that we're referencing each other. It begins and ends at the level of reference. Generally, my experience working with scientists has been, "Oh, let's have an artist illustrate or visualize a dataset." I'm not knocking that, but that already puts the artist in a box of having to think about an outcome that is completely uninteresting. Now, artists are guilty of doing that as well. It drives me crazy when artists reference science and take the authority of that science to somehow place authority on their own work.

So my transdisciplinarity tries to encompass what you're saying, I think. What if you let me participate in experiment design? That would be a stranger combination, because then there's something at stake for the scientist to say, "Why would I let an artist help design an experiment?" It doesn't make sense at first, but I still think that we should.

One of my favorite topics in message design is often asked only of scientists; What should we send if we made contact? And always, inevitably, it's math. That's sound reasoning, and I totally get why we should start there. But what if you let the artist choose which math? That gets weirder. My point is that we've become uncomfortable in really letting each other into our processes. We got stuck in a way of thinking. So that's my case for transdisciplinarity still having a place at the table.

**A. MEDITCH** I would go even further and say we don't let people in who are the public. They become an other. What if we let people participate, and do in the real world some of the art, or some of the science? What happens to their engagement with and understanding of the process?

I mean, I think about some of what you do, Geoffrey. It's simulation. Is there a way again to break it down, not just to discipline, but for people out there, who are engaged in a different way?   -39-

**B. TRACEY** On that point, I think another thing to consider is that with all these projects, there are a lot of people who are going to be needed to participate, but who won't actually, say, get to go on the mission, right? So let's assume that we have some magical physics that allows us to have an interstellar colony. Well, it's hard to imagine that we'll ever even hear back from those people in our lifetime. So I think it's interesting to see what kind of resources emerge. How do you muster that drive when most people aren't going to get to experience the product?

I mean, if you could muster enough capital, you could buy the necessary pieces to go. But we're not really at that point with Mars travel. We're just able to buy a rocket to get into low-Earth orbit, if you're not a human, but a robot.

So how do we figure out an alternate mechanism to muster enough "human capital" in order to make these things possible? When we move to interstellar travel, in most cases, the people doing it aren't going to see the results.

So I think that's an interesting piece to remember that most people working in these endeavors aren't going to be the ones going. It's an interesting piece to see that we have this global collaboration, but to what end? And how do you muster that drive?

**S. ROSS** So if we were to look forward, I wonder what the group would think of the culmination of the various things that are happening in the world today from a technology point of view. Think robots, general artificial intelligence, what's happening in data sets in computing, etc., and moving towards the evolution of the human species no longer being organic.

If it's no longer organic, and now it's a singularity situation, that all goes out the window, because now you can live forever, and you will be able to participate in interstellar space.

**G. WEST** Well, it depends on who "you" are. Easy to say "you," but—

**S. ROSS** That's true. Well, not you.

**G. WEST** No, but even "you" in terms of what you believe to be or what we conceive to be a human being, a species, and so on. There's also a kind of presupposition in that, that this new thing is somehow connected with us, that there's some even conscious connection. But we don't have, at least most of us don't have, a conscious connection to whoever came before us. I mean, we don't feel we're representative of what they were doing. So I'm not too sure. I think we should actually just see ourselves as agents in a grand evolutionary history, rather than identify with "you." The "you" is all of us, is all of life, basically, right?

I find that less interesting, because I don't feel connected to it. That's why I bring it up, actually. I'm much more interested in flesh and blood and what happens to flesh and blood, and the kind of consciousness we have. I'm very anthropocentric in that sense, and what's happening to us, what will happen to us.

I was particularly intrigued, because I really liked the idea and I've always resonated with it, that going out is to look in. And the idea of having huge horizons means maybe we won't get there. Maybe you'll never make it to Australia! And, in fact, I suspect hundreds of attempts were made by people who never made it. But it created something. And I think that some of these lofty goals will, in fact, help us understand how we are here and what goes on inside of us, our brains, and what happens to us collectively.

And, as you said that, I was thinking about Stewart Brand and the extraordinary insight he had to just show the whole Earth. I mean, that was amazing! NASA had the bloody picture but didn't realize quite what the impact would be. And it did have a big impact.

But, having said that, it hasn't galvanized human beings to recognize that we are one planet and that we're leading ourselves towards something that is quite dangerous. So we still need to somehow galvanize ourselves in a much more global fashion.

-41-

................................................................

## Wouldn't it be wonderful to instill the terror that an astronaut might feel as they're escaping even the view of Earth?

................................................................

**K. GREENE** Wouldn't it be wonderful to instill the terror that an astronaut might feel as they're escaping even the view of Earth? They say it haunts most people. I mean, going to Mars and just seeing Earth receding in your window—it seems to me to be just horrifying. I feel like if more people had the opportunity to see that and to feel the feelings that come along with that . . .

**G. WEST** Did you simulate that?

**K. GREENE** No, actually. They would prefer to keep a picture or a projection of Earth in the spacecraft, because they were afraid of what might happen to the astronauts who leave the view.

**G. WEST** Oh, that's interesting.

**D. KRAKAUER** Can I open this up to a few people who have been silent? You guys, on this side of the table, yes.

**D.A. WALLACH** Well, the point that Geoffrey and Kate just raised leads to a question. I'd be curious to see what compelling narrative could fall out of complexity science that would have the essence of a teleology in which people would really want to be a part.

So there's a sense in which people right now complain about the lack of a coherent sense of political ideology or mission; that's the project we're all sort of engaged in, collectively. The more I read the work you guys do and learn about the stuff that comes out of here, the question I always end up preoccupied with is, what teleology of information processing by the cosmos are we acting out? If we could understand what we're doing in our daily lives as small players, it would be very helpful for the people who are expounding this intellectual framework to suggest where it's going, or at least to suggest what our degrees of freedom are in imagining the different places it could go.

In the absence of any sort of sense of "Here's the mission, we're all in it together," I think people feel trivialized, and it's kind of the irony of things like the picture of the Earth. I mean, it's a great and interesting stimulant, but the irony of it is that it makes you feel trivial and marginal and unimportant and it sort of plays into a kind of nihilism that almost equally supports the project of burning everything down and being Epicurean.

**D. ROBLETO** Well, to be fair, that *Earthrise* photo did apparently spawn the modern environmental movement.

**S. MOORE-FABER** Absolutely.

**D.A. WALLACH** Yeah, it did.

**P. BUCCIERI** It worked very well.

**G. WEST** The *Whole Earth Catalog*!

**D. ROBLETO** So I don't want to start too far down that road of that image not mobilizing.

**G. WEST** It had huge impact.

**A. MEDITCH** Huge.

**G. WEST** But it sort of asymptoted somehow. It saturated. I think that's been one of our problems—that somehow, in some kind of sociopsychological way, it hasn't had a permanent imprint in terms of galvanizing us and responding to what are clearly potentially serious existential threats.

**D.A. WALLACH** Well, the net message you take from it is, "You're little. You don't matter." Not, "You're great. You're the only thing that matters, and the consequences of the choices are supremely important."

**A. KOHLER** One of the threads that's running through a lot of what I've heard here is, under what circumstances do we as a community globally do our best work? I thought it was interesting. To the very first comment Neal made, "We weren't able to terraform here, so how can we presume to be able to terraform elsewhere?" The first question I had was, well, did we do our best work? Was that a scenario in which we really fulfilled our full potential?

It seems like maybe InterPlanetary is a forum in which the group could postulate a bunch of necessary ingredients for how we do our best work. We know from history that having some sort of ideological motivation that brings us together, that galvanizes us, is a powerful force. We noticed that interdisciplinary work is something that seems to bring out the best ideas. And maybe there are many others.

But whatever the list is, maybe InterPlanetary can be a framework in which we can make some hypotheses around how humanity does its most creative, bold work, and then the story that could be told around this is, "This is a great opportunity for us to do that yet again. This is a great time for us to repeat something extraordinary."

**A. EZ** Doesn't it depend on who the target audience is? Because I think, arguably, to your point of not wanting to let the public in, I think maybe the discussion around the people in this room is different to a discussion that touches a mass population.

**S. ROSS** You think?

**K. GREENE** How do you upload it to the masses? How do you put it into language that everyone understands?

-43-

**A. MEDITCH** I was really struck by what you said, D.A., because it made me wonder if there's a generational difference here, that it was an incredibly powerful and galvanizing image. I'm really struck by the fact that you noted the nihilism of it, this feeling that this means we're really small. Because my first reaction to what Geoffrey said was, is there a way to reenergize that image? But you're saying something really different. How do we engage as a larger whole if we don't see ourselves as a whole? I don't know the answer to that, but, I mean, I'm really . . . the nihilism didn't strike me at all. So I'm really curious about what you said and what the implications are for engaging.

**D.A. WALLACH** Just to be clear, I'm not sure that it makes *me* depressed to see the image. I love it. But how you feel about it is also a reflection of how enjoyable your experience on this planet is. So if you're a global winner who enjoys the clean air you breathe and the city you live in and all the intellectuals you hang out with, of course it's something you want to hang onto and preserve. If you're in a more arduous existence, and feeling more anonymous, like the world's much bigger, like you have less power relative to the people who are taking the pictures of the planet from their spaceships, it's a little more alienating.

**G. WEST** Can I add something to that? Because I was . . . it's hard for me to actually empathize with this nihilistic view of this.

**D.A. WALLACH** Well—

**G. WEST** No, one second. I did see—in fact, Stewart Brand helped put it together—a three-minute video of the pictures taken from the International Space Station with music by Brian Eno, by the way. And this thing was extraordinary! Not just because of the beautiful images, but here's the thing that hits you: the whole fucking planet is alive with cities and people.

**S. MOORE-FABER** We have terraformed.

**G. WEST** We have. Exactly. I felt, "My God." You know, that sort of very primitive feeling of, "Who am I?" Like an ant or whatever.

That feeling isn't "nihilistic," I wouldn't quite use that word, but that feeling of numbers, of being swamped. So I did get that feeling as part of it, but it reinforced for me how crucial it is to understand that phenomenon. All that infrastructure, all that stuff going on, all of those people's lives and everybody trying to make money and win a game and so on, all of that stuff is going on down there.

**N. STEPHENSON** What one hears from people who've been up there is that the view from the ISS is just fantastic. The variety of both natural and human-built landscapes that pass under you at all times is absolutely mesmerizing and probably gives a whole different sense than the picture we're talking about.

-45-

**G. WEST** That's why I wanted to compare and contrast those two [reactions to seeing] the planet in its entirety.

**A. MEDITCH** So now you're talking about perspective. How do we give people the experience that they're having? I mean, how do you put other people there? That seems to me to be one of the things-

**N. STEPHENSON** Give them the ability to see—

**C. CONN** So what is the call to action, even if you do have the right message to rally people behind whatever the objective is?

**S. ROSS** I think it's even more than that. I think it's the acceptance of what that image might be. So I think you brought up a really important point of an intergenerational or generational difference. On one hand, we're listening to Joni Mitchell, and on the other hand you're listening to Rage Against the Machine. So there is a different sensibility of the way the different generations approach it. Now, I've been involved in virtual reality a lot, and virtual reality is now, if done right, which I'm not sure it is yet, could be seen as an empathy machine.

Nonny de la Peña, who's incredible at that, can put people into circumstances and situations that they would not normally be in and really feel it on an emotional level. I think, walking into it, you have to have the predisposition to want to feel that. If you walk into it thinking the world is dystopian and terrible, how do you move that person to feel that empathy? That "we are the world" feeling.

**P. BUCCIERI** When people are just trying to put food on their table, to feed their families, to just live, get through every single day, it's hard to dream beyond the street that you're in. My wife grew up in a town where there was extreme gun violence. When I was courting her, I felt fear going into that neighborhood and knowing that just last week somebody was shot down in the street. So it's hard to put people beyond those things.

So my feeling is that we are inherently selfish. First it's, "How do I feed myself, my family?" And then it's a huge jump to these wonderful things you're talking about.

**D. KRAKAUER** This is the charge. This is the discussion to have, right? I mean, this is the question that we're posing. InterPlanetary certainly isn't Burning Man. It certainly isn't Comic-Con. It certainly isn't a television show.

**P. BUCCIERI** But I love those, too.

**D. KRAKAUER** Yes, they're great, too. But they're not this. They all have exclusivity, for various reasons, not always by design. We're going to have a project, the InterPlanetary Project, which is extraordinarily ambitious. And what you've said, you know, these remarks about feeling included, or being teeny, or making a marginal contribution, the mundane reality of day-to-day life, etc., is true. And yet, there is this huge appetite for what your TV companies have done and what Neal's books do, and what other people in this room do. So what should we design that recognizes those limitations and that is not being addressed by what we have currently? Is it impossible? I doubt it. I mean, that is exactly what we're here for.

**P. BUCCIERI** I totally get that; how do we create access? So virtual reality is a classic example, but a lot of people never experience that.

**S. MOORE-FABER** I'd like to distinguish between the luxury of thinking about these problems today as a very elitist situation. I mean, our group here is not representative at all, but people could have access to the ideas and enjoyment of the ideas that are produced here. I'm much more hopeful about the second part because, you know, movies and culture reach so many people on the globe. Ultimately, I think we can be inclusive, but not necessarily at the beginning.

**P. BUCCIERI** I agree.

**A. EZ** But if we take what Paul was saying and apply the rationale that Neal started off with, don't you require an external threat in order to motivate people?

**P. BUCCIERI** Yeah, fear. It's a great motivator.

**G. WEST** Yes.

**A. MEDITCH** We have to agree that there already is a threat.

**A. EZ** Well, this is where it gets tricky, because the threat is this abstract thing that is difficult to grasp. It's not the same as a bunch of dudes with pitchforks who are chasing you out of town. I think that's where there's this kind of epistemological crisis at the moment, in that we can't quite explain what this thing is that's probably destroying us. I'm going to use a *Game of Thrones* analogy, because we've managed to get about forty-five minutes into this and, astoundingly, no one has mentioned it, yet.

LAUGHTER

**D. KRAKAUER** I think I might have mentioned George R.R. Martin in the very beginning.

**A. EZ** It's like there are the White Walkers beyond the Wall and nobody south of the Wall except for Jon Snow really gives a fuck. It's this kind of inability to communicate things that will have dire consequences but aren't summed up easily.

**D. ROBLETO** So I want to challenge Paul's point just a little.

**P. BUCCIERI** Challenge away.

**D. ROBLETO** Okay. In my experience, arts education is a big part of my practice, and specifically with underrepresented youth, the economically deprived. You bring art in, it's there. It doesn't take a lot of work. That's why I know scientists will feel some sympathy to my deep worry about the lack of arts education in our country.

I can get to any profound point we're talking about at this table today on day one if you allow someone the access, so that their voice, as a creative voice, is valid right away. My own experience has generally been that art is an incredible educational tool to allow for that.

The other thing I'd like to throw out on the table, in our criticisms of how messages may or may not have succeeded to date—to me, we have to define what we mean by success in terms of numbers of people who bought into the message. Because was *Cosmos* an unsuccessful television series? Absolutely not. Are the ones you're working on unsuccessful?

**D. KRAKAUER** What is success here? The number of viewers, or actual impact on the mind?

**D. ROBLETO** Well, that's what I don't understand. Are we being hard on ourselves because the whole planet hasn't mobilized? What could happen that would perhaps mobilize us in this idealistic way that is a global movement? It's a pretty short list. I do think contact with another intelligent life form may be one of those things. I can already sort of predict where that discussion will quickly go off the rails.

But why are we saying what's been done to date didn't succeed?

**J. FLACK** I think it's worth pointing out that the two distinctions I'm about to lay out on the table are not really strongly separated if you think hard about them. But they are distinctions.

So Geoffrey said that the Earth is terraformed. Well, that's true, but it was terraformed over very, very long timescales organically. And any mission, for example, to Mars, where we start thinking about how to terraform Mars, physically, ecologically, and socially, is really an exercise in quantitative whole-planet, large-scale engineering. It's a totally different thing.

I don't think we actually need any sort of outside threat. I don't think we need to get people involved in the experience through VR. I think the number of people that is going to be needed to make that project happen is already vast. It just requires an incredible discussion on the ground about what constitutes a well-designed society. What kind of democracy do we want? Does democracy scale? These are all questions that one is going to have to ask in this project. And they're not just questions for elite people sitting at this table. They're questions for everyone. That's how people are going to get engaged.

-49-

**T. ASHCRAFT** Personally, I wouldn't mind transcending my humanity and merging my mind with a greater organism. I work towards that. There's an urge towards that. That seems like dynamic territory. You know? Sometimes it tells me there's something called the biological commonwealth, political state, whatever that is. And from my tripping, I've seen that, I've gone into that. It seems like an infinite place and it makes sense.

I have one other thought that relates to the earlier conversation about going to Mars. I sometimes think rather than go, which would be fun, can we pull destinations to ourselves? I just keep that thought in mind. That guides me.

**D. KRAKAUER** To Thomas's point, this is all very "techie." And Jess's point, I think, is very well taken, which is in this country at the moment—and not only in this country—there's a lot of conversation about alternative political realities, given the abject failure of some of the institutions that exist, in terms of representing many people, etc. I don't need to go into it in depth.

One of the issues, and certainly some writers—our friend Kim Stanley Robinson is a good example—have used space as an opportunity to explore not just rockets or telescopes or scientific issues but alternative political realities. That's something where the barrier to entry is lower. In other words, you're an activist, you could say, "I have an opinion on how society should be structured or restructured." If you have kind of a *tabula rasa*, if you said, "I could start again," how would you start again?

I think about economic systems, too. Charlie Stross wrote a lot about economic systems. So that's something that should also be in this conversation. It shouldn't just be the sciency stuff, and it shouldn't just be the arty stuff either. It should actually be issues about the mutability of human society itself.

**J. FLACK** And the design. This discussion can be aided in a way that we were never able to do before, because we didn't have the data. We didn't have the microscopic data on individual interactions on how people get along, and we didn't have AI to do pattern recognition. So there's an opportunity to do this large-scale engineering in a quantitative way that we have never had before. In that way, it's incredibly exciting, right?

**G. WEST** We could actually think of the experiments that people have done for centuries, of trying to form utopian communities. I mean, I can speak personally. The reason I originally came to New Mexico is because northern New Mexico was a hotbed of communes and so on. Much of my time in the '60s and up through the '70s was spent ruminating about different forms in which people could live together in terms of the communal structure, in terms of individual relationships, the idea of marriage, and children, and of the economy, shared economies, and so on. It's not so different. It's not that people haven't been doing this. Jess, you're suggesting that now we can do it better, maybe, because we may understand the science of interaction. I would love to believe that. But I'm somewhat skeptical.

**N. STEPHENSON** I can recommend Cory Doctorow's new book, *Walkaway*, which has been out for a couple of months. Its original title was "Utopia," but they made him change it, I guess, to *Walkaway*. It ties together what Brendan was talking about, the kind of open-source ethos, you know, what if we did an entire alternate civilization and culture based on the way that open-source communities work? It ties in 3D printers and GitHub, and it's a very carefully thought-out depiction of how it might actually work.

**A. MEDITCH** I love what you just said, it was so beautifully clear to me. I mean, "If you could start over, how would you design, or how would you make, a world?" I think there's a way to draw people who are alienated into that process.

## If you could start over, how would you design, or how would you make, a world?

-51-

**C. CONN** Why are we taking this into space? I mean, to Neal's point last night, like, why don't we try that on Antarctica instead? With all of these great ideas and this sort of utopian ideology, and the new information we have, wouldn't it make sense to try that somewhere on our planet? Or are we already entrenched?

**N. STEPHENSON** Well, this is what the story of *Walkaway* is. It's about people literally walking away from a society that is dominated by hyper-wealthy elites. In this case they're wandering off into Canada. But it could be—

**S. MOORE-FABER** I wonder why.

**N. STEPHENSON** It could be anywhere. It's essentially a viral system of building the kinds of communities that everyone here has been talking about for the last few minutes, that are open to people who aren't necessarily members of a moneyed elite. Heartily recommended.

**A. EZ** Does it come out with an optimistic view of that form of living?

**N. STEPHENSON** I would say it's not optimistic in a kind of naïve way. I would say it's optimistic in a pretty nuanced and sophisticated way.

**C. KEMPES** So, in addition to what you said about this being an area where you can bring in lots of people and have lots of participation, I think it's also a place where you can convince scientists that there are relatively understudied and interesting problems. So you can walk into any elementary school classroom and you say, "What's so hard about going to Mars?" They'll give you a billion answers. And if you say, "What's so hard about distributing food in America?" . . . Even to undergrad or PhD students in science, that's a much harder question. And so I think that part of this is communicating that there are also very hard social problems that we have less of a language for, and less of an ability to look at.

**D. KRAKAUER** Please, can I just push on that? There's something here we're just fascinated with. This discussion tonight is completely sold out, by the way. Why? Now obviously it's because of all of you extraordinary people. But also because people are extraordinarily interested in this. In InterPlanetary. And, in this way, I think that it's partly suspicious and partly optimistic. But why is it that that language is so effortless? Say you ask in high school, "How are we going to get food across the surface of Mars?" And students immediately start thinking about it. But you pose it as an economics question, and they'll go to sleep in a millisecond. So I'm curious about this: What makes posing questions in these terms so much more salient and motivating than the way we do in our education system, which essentially turns people off?

**C. KEMPES** I think to answer that question I would want to talk to both psychologists and people who understand storytelling. Because the hypotheses that come to mind are either humans are good at thinking about logistics and so we start thinking about technology and planning, we're good at that. Or that we've watched many of these types of stories growing up, right? We've seen many space stories. They've given us a vision into another world. If you make any story in an interesting way, it gains momentum. I don't know, maybe if we had had fifty years of very interesting, alternate perspectives on societies, this would be a more natural way of thinking.

**A. MEDITCH** But we're using story in a way that is still talking about pushing something out. I think part of what's being talked about is the emotional engagement when you can participate, when you feel like you have a stake. Where you're part of the making of a story.

**S. ROSS** A question for you, David. It seems to me that the conversation is leaning away from InterPlanetary space and towards trying to solve issues and problems that we face on our own planet. Are you okay with that?

**D. KRAKAUER** Yes! I'm okay with anything that this room does.

**S. ROSS** Because the marquee said "InterPlanetary" . . .

LAUGHTER

**D. KRAKAUER** I know, I know. It's interesting, because what this shows is the gravitational well, at least amongst this demographic, is just that one, our planet. But let me push things a little bit away now to be more specific, which is to talk about what we plan on doing. I'd like to get your thoughts on this. Part of why you're here is not just for this panel discussion, but so we can think about implementing the InterPlanetary Project, right? There are many objectives. I don't know, Caitlin, if you want to sort of remind people of the schedule of events? I really do want to talk a little bit now about what next year will look like, and some of the things we're planning on doing. Because how do we get some of these issues and objections and concerns out there into something very tangible, something that we're going to build? What should its elements be?

**CAITLIN MCSHEA** Yeah, sure. So, following on the heels of tonight's first InterPlanetary discussion, there will be a five-day series of events in mid-October. On October 13, we're having our follow-up to this panel discussion, so there's a lot of room to figure out what we want to tackle there and then. That Saturday is a city-wide sci-fi film festival, and we'll be showing really interesting and

-53-

relevant movies at all five of the independent theaters throughout Santa Fe. On Sunday, at SITE Santa Fe, we'll have an event at *Future Shock*, which is a fantastic show in which Dario is one of the featured artists. It's also the grand reopening of SITE Santa Fe. They will be hosting "InterPlanetary Pass Holders" for a morning lecture that Geoffrey will be giving. We'll have a book signing there, too, so it's a really wonderful opportunity to see the *Future Shock* exhibition and to hear Geoffrey discuss *Scale*. Dario may or may not be there, we'll see. And then, on Monday, we're collaborating with MAKE Santa Fe for a citywide InterPlanetary "Space-Craft" event. So creators and artists will be demonstrating and showcasing their work at the Maker Space there. People can learn how to use these printers and machines. And then, that Tuesday is our Anthropocene Community Lecture featuring Manfred Laublicher and other panelists, TBD.

So that's the outline for what's happening in October, and perhaps there might be something between October and June. But the big festival, the inaugural InterPlanetary Festival, will take place June 7 and June 8 in the Railyard. There will be sci-fi film screenings taking place there, too. There will be accessible Maker Spaces again, live concerts, gaming. There's food and beer. It coincides with CURRENTS, a really wonderful annual immersive new media installation here in Santa Fe. There will be lectures and panel discussions throughout. But all of this fun stuff will be grounded at its epicenter by an Innovations Expo, or a space-technology expo. We're inviting leaders in technology—the people who are already working on interplanetary exploration—to come and get a little booth, a very modest booth, to show off the coolest things they have with regard to the technology that exists, or with regard to the technology that could exist. These organizations are all going to be showing their stuff next to each other. It's really an opportunity for the community to engage with tangible examples of what's already available. And, since this will take place annually, we will see in real time how these technologies develop. Eventually, maybe fifty or a hundred years from now, there might be a satellite festival on Mars, who knows? But that's the general framework, so if you guys want to take that and run with it, go for it.

**D. KRAKAUER** That's great, and this is something that Scott and I talked about on the phone—the World's Fair and what the World's Fair meant to many growing up, and the inspiration it had on you. And what is a World's Fair in the twenty-first century with a population of this many people with this massive economic inequality? What is access to a festival given that a festival takes place in one place? I'd sort of like to have that conversation. What should we be building? We are going to do this, by the way, we're going to do this. So what should we add to it that will ensure that it has an impact that's greater than just a few hundred people being entertained for two days?

**G. WEST** Did we discuss what we're actually trying to accomplish with this? This is a very incipient stage, actually. But what is it we're trying to do? What is the message?

**D. KRAKAUER** It's pretty straightforward. I think it's something new, which is developing a genuine intuition for what a living planet is. A living planet has cities, and it has farms, and it has technology, and it has religion, and it has science. It's a very ambitious project. I think our sense is, the sense from everyone in this room, is that the world is so connected, effects propagate quickly, and the population is growing so fast that, unless we develop a global sensibility, all of these little fractionated, disciplined forms of knowledge are going to be insufficient. So that's the big picture: global intuition. How do you develop it in a way that's exciting? So you would think, oh, there's an audience, but also, it's potentially changing people's lives. It's *that* ambitious.

**J. FLACK** There's another major difference. And that is we start here from the local and we project to the global to develop that intuition. But the target there will be starting with the whole planet, actually, and projecting down locally. That's the fundamental framework.

**G. WEST** How does that, what you just articulated, relate to whatever the concept is of InterPlanetary?

**A. MEDITCH** The tagline made it absolutely clear—I mean, to your point, it's one planet at a time. And we kind of have to start with this one. So to me the implication is right there for us to play with in terms of the messaging. I don't know what the other communications folks think here. I would also add another question that I have. Who's the audience this first time out? Is it other people, clones of us? Is it people on the south side of Santa Fe? Do we have a sense of that yet?

**D. KRAKAUER** We've talked about that briefly before. Do we have a market? We know people are interested in what we're doing because we have magazines like *Wired*, sites like The Verge, Ars Technica, io9. That whole demographic is immediately in line with this, because they love it, right? You could argue, in that sense, if you're asking about market, the market exists. What seems to me the more challenging thing is what's really come up in this room, which is a real spirit of inclusivity. I find that a much more interesting issue. I think the first thing is what guarantees the viability. The second thing is what motivates it, and how we do that?

**D.A. WALLACH** We need Breitbart involved.

LAUGHTER

**D. ROBLETO** On a nuts-and-bolts level, I strongly believe that in order to become an interplanetary society we have to rethink our educational system as it relates to the arts or, more broadly, creativity research. I'll just be really broad. And then neuroscience comes into play here. The creativity research of neuroscience—

**D. KRAKAUER** I don't want to derail this question, but in terms of the language you're using, Dario, it's a language I try to avoid. It's sort of "arts," or it's "neuroscience," etc. We have this tendency to recruit this language and I think that's how most people talk about the world. They carve it up this way, and it's outrageously boring. Let's be clear—

**D. ROBLETO** I think you have to justify that more.

**D. KRAKAUER** I think we're sick of it.

**D. ROBLETO** Who's sick of it?

**D. KRAKAUER** I am.

LAUGHTER

**D. KRAKAUER** I think there's a way in which we're presenting a state of knowledge on the planet Earth that's fifty years out of date. And I think that when you talk about arts, it very quickly sounds like advocacy. "I want to support the arts," and so forth. That's not where we want to go. We keep drawing on this language to support the project. I don't even like this language of saying "the arts" and "the sciences." I actually feel that one thing that InterPlanetary could do is not do that. I'm much happier with your language of creativity.

-57-

**D. ROBLETO** But that seems to be a different battle. So you want me to invent a new word so that we can all focus our attention on the concept of art? That's crazy.

**D. KRAKAUER** No, I'm not saying that. I'm saying that there are challenges that are of extraordinary interest. And there are enterprises that extraordinarily creative people engage with. And I don't think it helps very much by using these meta-labels.

**D. ROBLETO** I think it would be important to know that we're using language that's very personalized, and that we might not all agree with. I didn't realize that *art* and *neuroscience* are now red-flag words for you. As leader of the initiative, we should maybe know that.

**D. KRAKAUER** No, Dario, that's not what I mean. I mean that it would be a shame to push that language forward again. So you won't see it at the Santa Fe Institute, you won't hear it there, and you wouldn't have heard it for thirty years. And yet you would be doing work that spanned them. So we found a way to work and collaborate without referring back to what look like disciplinary affiliations. You don't need them.

**S. MOORE-FABER** A slight change in tack here. If we're looking for what the mission is, I'd put forward the title—I wish I'd thought of it myself—of Tim Ferris's book *Coming of Age in the Milky Way*. I think that's our project, and what does that mean? We don't even know what that means yet, but that's why I'm here. And as part of that big subject hasn't been mentioned by this group, most of the discussion that I've heard is premised on human beings as they are now. Whereas it's very obvious that we're starting to tinker with our genome, and where that's going to lead us is a major, major question. It raises practical questions and moral questions, and I think there are many moral questions that are bound up with the project that we're talking about now, not the least of which is, what are we trying to accomplish and why do we think it's important? That's perhaps the biggest question.

-58-

............................................................

What does an intelligent species look like and how do we harmonize the motivations of the individual members of the species to be consummate with that larger goal?

............................................................

I guess what I'm trying to say is that I'd like to add the essence of intelligent species as part of our question. I think that's something we need to examine. And I think we need to examine, assuming that we want to be at home in the Milky Way, what does an intelligent species look like and how do we harmonize the motivations of the individual members of the species to be consummate with that larger goal? And right now I submit to you, I don't see that they're very consummate. Our big problem, it seems to me, is that individually and as a species we just want to consume more and more. I don't think that's a good recipe for success in the short term. But the irony for me is that, in the long-term, it's that kind of impetus that will push us to the stars. So I see great contradictions between the short term and the long-term, and that's what I'm wrestling with.

**D.A. WALLACH** So I was joking about Breitbart specifically—

LAUGHTER

**D.A. WALLACH** —but the point that I would seriously put forward is that there's a risk of this becoming a very pat, inspiring Spaceship-Earth, lofty thing, and everyone knows that aesthetic intimately. And if part of the goal is to get people juiced up about some missionary way of thinking about this, it may be productive for us to create some very provocative and uncomfortable stunts that are a part of this. So as opposed to, for example, proposing utopian visions of interplanetary civilization, what if we had a contest for the most repugnant proposals about how to organize systems or people or substantive outcomes in that civilization? What are the things we can do that would really make people uncomfortable, get people angry, or get people to come to Santa Fe to protest what they thought this was about? How can a performance art element be added to the entire thing so that it's a bit ironic and subversive?

-59-

**P. BUCCIERI** May I ask a question? What does "big" mean? Like, how many people is big? What's the view of that in terms of success? I see all the offshoots there.

**D. KRAKAUER** In terms of the physical footprint of the festival?

**P. BUCCIERI** I'm now just honing in on the festival.

**D. KRAKAUER** Then, I assume we're talking about in this town. I don't know, Caitlin, if you have a sense of what we can allow in this space, but we're talking several, is it tens of, thousands?

**C. MCSHEA** No, I don't think so.

**D. KRAKAUER** Not even that?

**C. MCSHEA** I think the physical limit is maybe twenty thousand.

**D. KRAKAUER** But there is a physical limit here.

**C. MCSHEA** Yes.

**P. BUCCIERI** I was just writing down travel, cars, hotels. I love this idea, by the way. I actually think, with this topic, you could really, really, really make it big. We just did a silly little topic for a show we called *Ancient Aliens* that's on the History Channel; we did a thing called AlienCon with very little promotion. We were not prepared for the outpouring of people. It just blew up to be so big we couldn't accommodate everybody.

**G. WEST** Where was it held, by the way?

**P. BUCCIERI** In San Diego. So we had to open up the small ballrooms. We had to open up the entire floors. So this thing to me, with this group of people, I think could fill this entire town.

**C. CONN** I think there's another element that we haven't talked about, which is why Santa Fe? Why New Mexico? And so part of the premise of this whole concept was of course to start this in Santa Fe, have this be sort of the hub, the vortex where everything happens. But we can certainly talk about this spreading throughout Roswell, the VLA, White Sands, Chaco Canyon, all of those. Everything about New Mexico is why this particular topic is so resonant here. And so I think we look at starting this in Santa Fe, but then as we build and plan and think about what this could mean. We could create these dystopian societies throughout New Mexico.

**D. KRAKAUER** Yeah, and you can rule over them!

**K. GREENE** What about efforts to reach out to minority populations? So, for example, Afrofuturism. I would love a panel on Afrofuturism. These things have been talked about in communities that have not at all been represented here for a really long time. And those ideas aren't legitimate until they become popular and commercialized. And I think there is huge space for that in this discussion.

I think about the poets of color that I know. And the stuff that they're doing. They're thinking about identity and how we could all be together, the same. These are discussions that are not compartmentalized for them. They're everywhere in their lives and bodies and psyches.

**D. KRAKAUER** How do you make that appealing to millions? In other words, the goal here is to show the extraordinary quality in that. It's like our science, right? At the moment, it appears inaccessible, forbidding. And again, some radical poetry would appear political or polarizing. What do we do to take all of this extraordinary quality and demonstrate it to be just that at an extraordinary scale? That's the goal.

-61-

**K. GREENE** I think SFI has the ability to legitimize almost anything. So if you say this is important, people will love it. They'll be curious. So I think the authority invested in white people and people in positions of power to say, "Pay attention to this," is undervalued, and I think it could be used here to some great effect.

**M. TOSH** Aren't you all essentially doing this tonight? You could easily have this discussion there. You've opened up the Lensic [Performing Arts Center], and you're going to have totally different faces there tonight than normally come to the sort of events you guys put on. Isn't beer allowed? It's BYOB, right?

**C. MCSHEA** It's not BYOB. It's "buy in the lobby."

**S. MOORE-FABER** BITL.

**M. TOSH** That works, too! But that's something completely different in a way. You're thinking about the science, and then merging it with people who have the ambition to care enough to even ask a single question. I think tonight, you guys are going to be on a panel, and you're opening up the floor to people who are going to ask some pretty crazy questions, I can imagine, given that we're in Santa Fe. That's going to be test one, you know? We get to witness it unfold.

**S. ROSS** To get to June next year, so it takes place in Santa Fe, and let's say there are twenty thousand people, and there are these events and things that happen, but that won't reach a global audience. In reaching a global audience, I'm going to use the word—I hope I don't offend—"transmedia." You have to be able to cross every form of media possible and have, as we talked about, a virtual, digital World's Fair that plays throughout the planet. And then I think if you bring in the necessary sponsorship and people and support to publicize it, it could be that big.

**G. WEST** I love the idea of a virtual World's Fair. That's a great idea, actually. Lots of virtual reality.

**S. ROSS** Well, the big problem there becomes, how does one view it? So right now we're still caught up in this "Is it cardboard? Is it Samsung VR? Is it Oculus?" debate. Hopefully, standards get to be issued.

**D. KRAKAUER** So one thing for everyone to be thinking about is that we do have a very tangible road map. I think there are things we're really going to do and we can all imagine in our heads what some of this would look like. But what's different here is that there are issues of inclusion. That came up a lot, which is interesting. But it's also the fact that this is rigorously grounded. And that is not true with any of these other things. It's certainly not true with TED. TED is just kind of shallow, intellectual eclectica. And that's fine; it'll die in the next two years.

**S. ROSS** Can I quote you on that?

**D. KRAKAUER** Yes, you can quote me on that. I've said it many times. I can't stand it. And I think that if it evaporates from the face of the earth, oh well! And so the question is, how do you get to the real ideas? I think people have actual appetites for real ideas.

**L. WALLACH** They have appetites, but they don't have the access. When I think of TED, TED really grew when, all of a sudden, people got invited to something they wouldn't be otherwise. Then came YouTube. Now it's like, "Okay, this is not an official TED thing, but we'll let you use our name," and they started these regional TEDx things. So it was giving people who don't otherwise have the means of transportation access. It's all about access. People are curious, but if there's nowhere to apply it, curiosity just dies. I give TED a tremendous amount of credit for keeping a variety of really interesting ideas and relationships alive in the lives of people who otherwise would never have an opportunity to engage.

**K. GREENE** Is there the ability to bring in people from JAXA or the Indian Space Agency or the UAE, which is really pro-space? It's spending millions upon millions of dollars on their space education program, just to jump-start education, in general. Getting people off their petrodollars and onto another path. And I think that really gives it a global sense. People are working on the problems of space, that's interplanetary in a way. But there are really different ideas about how that comes about. Just briefly, in India, they get projects done a lot faster, a lot cheaper. I wrote a story about it recently, and I learned that there's a lot of bypassing bureaucracy that's happening. So it's not a technology problem *per se*—their technology is probably about thirty [years] behind the stuff that Russia and the US use—but, cleverly, they use a slingshot orbit to get their probe to Mars. And they bypass the usual hierarchies to get a project off the ground and running, and then launched. So there's a lot to learn from that, I think. And then also what the UAE is doing, culturally. They're trying to create a culture where space exploration is aspirational.

**D. KRAKAUER** That's certainly something I want to discuss tonight: the economics and the differences of $60 million in India and $600 million in the US, because I think it's a really interesting discussion. One thing I'm not sure came across when Caitlin was describing it, but just so you understand, this festival is taking place in an area that has a farmer's market. So those little booths that you normally associate with fruits and vegetables and popcorn and so forth are actually populated by companies, whether it's Electronic Arts or Blue Origin or what have you. And that's what they're going to do. And they're going to have to bring their A game to this. We're not asking for brochures. They're not hiring people. It's about showcasing what makes your company or organization interesting and creative. And that's the sponsorship model, too, for those companies to come and contribute money to actually be there and present to this audience their vision of the future. And so that's something to bear in mind. So when you say the Emirates, absolutely we should have their space agency participate in that first.

**D. ROBLETO** So the rigorous part you brought up as a standard for this, you mean that it's grounded in rigorous science? Like, people who would be accepted for some of those booths, it's not a crop circle booth?

**D. KRAKAUER** No.

LAUGHTER

**D. ROBLETO** Okay, but is that what you mean?

**D. KRAKAUER** Yes, that's what I mean. I'm not interested in that. I think there's enough of that already, so we don't need to have that here.

**D. ROBLETO** Because I feel a tension here, if I may be honest, if that is to work at this kind of scale, you're going to have to have some comfort with the populist side of this. And you've voiced your suspicion about certain things and language and motivations behind them, but there is going to have to be some room for that.

**D. KRAKAUER** So we have Cormac McCarthy here with us today. He writes very good books, admired by many people. Jonah Nolan couldn't make it here today, but Jonah wrote *Person of Interest*, *Interstellar*, and *Westworld*, all fantastic. If we can get to an audience that big, we're happy. And there's no trash in it. So I would make a distinction between really great populism versus a kind of desperate pandering. And, you know, Neal is sitting right next to me—he has extraordinary numbers of readers and he writes brilliant books. I think we can be extraordinarily popular and be great.

**D. ROBLETO** But it just makes us the judges of where that line is.

**D. KRAKAUER** That's fine.

**D. ROBLETO** Okay.

**A. MEDITCH** We haven't yet talked about the nuts and bolts about the festival, at all. Given a lot of what's been talked about, I've just been eyeballing *Walkaway*, the review on NPR, and this idea of, If we start over, what does it look like? What are the systems you would like? What would work? What wouldn't work? What's the worst example of this? That's something that could be designed as a participatory exercise, from a kid in grade school to a scientist working in a related field. I would urge there be an interactive component that would engage people and give them buy-in.

**D. KRAKAUER** Of course, absolutely.

**A. MEDITCH** I was even thinking tonight at the panel, we can't do it because it's tonight, but you could pass out cards and have people write one thing on it. What's the one thing that they think is most important in a well-functioning society? And what's the worst thing?

**G. WEST** There are projects where this is now being done practically. We're actually designing new developments in cities and so on which are totally engaging with the people who are living there, or will be living there, and so forth. And the whole design evolves in a kind of collective, crowdsourced way.

THUNDER RUMBLES

**D. KRAKAUER** So the weather is telling us it's wrap-up time. And I think it's fine that we have disagreements. There should be tension in this room. And I think we are to work toward something really interesting. So I really appreciate it. I know you're busy folks. I think tonight's event will be fantastic, and let's keep this going somehow. I'm not quite sure how, but let's keep building momentum gradually through October through to June and beyond that. Thanks very much.

**G. WEST** Very good!

**D. KRAKAUER** It's going to be an interesting and rocky journey.

Proceedings of
Santa Fe Institute's first
INTERPLANETARY
FESTIVAL

JUNE 7–8, 2018  |  SANTA FE, NEW MEXICO

# INTRODUCTION:
## PLANETARY POLICY & REGULATION

Can a deeper appreciation for complexity usefully inform policy?

Certainly, one hopes so. The most critical problems of our time—from water, agriculture, and climate change to the design of cities, economic development, and the stability of the financial system—involve higher levels of complexity than the traditional systems of law and policy have been built to manage.

As a result, *complexity* and *systems change* are becoming favored approaches to envisioning new policy-based approaches to these and other issues.

Yet, as Geoffrey West notes in his new book, *Scale*, "Existing strategies have, to a large extent, failed to come to terms with an essential feature of the long-term sustainability challenge embodied in the paradigm of complex adaptive systems; namely, the pervasive interconnectedness and interdependency of energy, resources, and environmental, ecological, economic, social, and political systems . . . It's time to recognize that a broad, multidisciplinary, multi-institutional, multinational initiative, guided by a broader, more integrated and unified perspective, should be playing a central role in guiding our scientific agenda . . . and informing policy."

The InterPlanetary Festival is a step in that direction. Discussions inspired many in attendance to think about problems here on Earth, right now, while also focusing on an interplanetary perspective and thinking about long time horizons, centralization and decentralization, rights, values, and risks.

As Linda Sheehan noted in her remarks, "What we're learning from the work done at the Santa Fe Institute about complexity is really informative to law."

One set of insights concerns how complexity plays out over time. Modern markets and legal systems are poorly suited to support responsible consideration of the needs of future generations, yet as Sheehan noted, "indigenous customary law generally focuses on both ancestors and on what is owed to the ancestors, the people who came before us, but then also looking ahead to the proverbial seven generations. How are we looking ahead?"

-69-

Complexity also plays out over *space*—the discussion of interplanetary law and policy highlighted the tension between local choice and governance and universal concepts that imply some centralized authority or single culture. A rather optimistic take on the possibilities associated with bringing new ideas about complexity and scale into policy making is that there may be rights and "laws" that are universal but not centralized.

> ...............................................................
>
> ## What we're learning from the work done at the Santa Fe Institute about complexity is really informative to law.
>
> ...............................................................

What if ideas about *complexity* as an organizing concept are poised not so much for broad, constructive, and beneficial adoption as misuse and misapplication? For example, as justification of irresponsible governance (the financial system is a complex system that can't be regulated closely) or authoritarian control (finally, we understand how to manage society scientifically)?

The latest concepts in science have often been appropriated to support self-serving arguments by those in power; it's easy to imagine a book like Donella Meadows's *Thinking in Systems* being adopted by technocratic elites to justify various top-down, coercive interventions for the good of the governed, much as the *Origin of Species* was adopted by plutocrats and social Darwinists of the late nineteenth century to justify the social order of that time. Given the modern crisis of legitimacy facing those now making policy, borrowing a little scientific credibility has to be tempting: What's more legitimate than "science"?

> If legal systems reflect the complexity of the societies they govern, and if interplanetary society is the human future, then the prospect of legal systems and policies too complex for human understanding is one that deserves consideration now.

The application of new abstract principles onto messy realities may come from the best intentions. But if a scientism of complexity is not to become more influential than the science of complexity, SFI, and those associated with it, will need to work hard to ensure that the genuine insights from their work—an awareness of unintended consequences and not-yet-obvious connections—are integrated into policy rather than misappropriated or applied unwisely.

Still, if legal systems reflect the complexity of the societies they govern, and if interplanetary society is the human future, then the prospect of legal systems and policies too complex for human understanding deserves consideration now. Additional thought experiments about the possibility of beneficent interplanetary regulation might at least help us anticipate some of the troubles—and potential advances—of tomorrow.

What's needed, as SFI President David Krakauer noted in his introduction to this panel, is an appreciation of "how regulatory systems interact in such a way as to promote freedom but also good behavior and long-term vision."

-71-

*—Jeff Ubois*
*Senior Program Officer*
*MacArthur Foundation*[1]

---

[1] The views expressed here by Jeff Ubois are his own, and do not represent the views of the MacArthur Foundation.

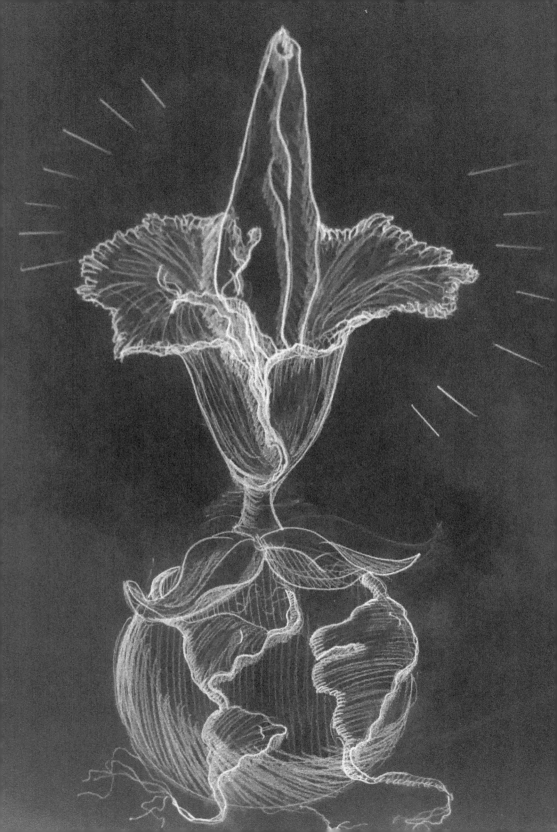

# PLANETARY POLICY & REGULATION

*David Krakauer introduces the panel,*
*featuring Linda Sheehan and Jeff Ubois.*

**DAVID KRAKAUER** We've been listening to issues around creating sustainable planets, and one such issue is human beings interacting with their environments and with each other, and that's a question of regulation and law. So I'm going to introduce the two discussants who are going to be talking about those dimensions of this InterPlanetary experiment.

Linda Sheehan is senior counsel with the Leonardo DiCaprio Foundation. And Linda is trained as a lawyer and is particularly interested in what I would call in our conversations *complex law*—how regulatory systems interact in such a way as to promote freedom but also good behavior and long-term vision, and that's sort of Linda's issue. Of course, one of the most pressing problems in that space is climate change.

I shouldn't really need this cheat sheet for Jeff, because I've known Jeff for a long time. Jeff works at the MacArthur Foundation, which has been supporting incredible research for a very long time. And Jeff runs the 100&Change program—I'll allow him to talk about that a little bit. But he's also something of a social activist. I think he brings to foundations the spirit of the open source movement, the idea that you need to create tools to empower citizens, not to diminish their freedoms, and so I've always been a great admirer of the work that he's been doing. So with that, Linda and Jeff.

APPLAUSE

**JEFF UBOIS** Great. Thank you, David.

**LINDA SHEEHAN** Thank you.

**J. UBOIS** Welcome! Thanks for coming out in the hot sun. If you look at the session's description, you'll see that we are going to cover a lot of topics in a short time, and we'll be looking for dialogue following some of these remarks. When Linda and I were talking about this panel, we both noted that often it's easier to understand the principles regarding Earth by thinking about places far away. And so, as we get going, we encourage you to think alternately about places far away and places here on Earth, where it does seem that many of the problems that are facing us now reflect some kind of crisis of complexity. If you think about climate or economics, or health care, or demographics, these all involve higher levels of complexity than the traditional systems of law and policy have been built to manage.

**L. SHEEHAN** As the lawyer here, I feel a little bit uncool. I almost feel like I'm shutting down the party. But nobody's going to get in trouble! We're going to talk about what our legal systems look like now and, as we look out toward space, what that might teach us about what we want to do going forward both in space and right here on Earth. Now, both of our foundations fund extensively in the area of environmental conservation, protecting the systems that we have right now. I think perhaps Jeff and I might start out by setting the stage in terms of the context that we face and that we see our grantees facing on a regular basis. I went to MIT undergrad, and I like to go back to MIT's John Sterman, who talks about sustainability, runs the system dynamics program at MIT, and he summed it up for us nicely. He said that our civilization is unsustainable and it's getting worse fast. That pretty much gets to the point! He says that the trend is for more of the same, and more quickly. These have real significant impacts at home. I tend to focus on water issues. Over a billion and a half people in the world today are now living by river systems that are using more water, more than even the minimum rate of replenishment. We're going to be facing some serious water shortages. And it's not like we're talking about chia pets—maybe I'm dating myself here—we're not. They're humans, and humans need water in order to survive.

.................................................................................................

Over a billion and a half people in the world today are now living by river systems that are using more water, more than even the minimum rate of replenishment. We're going to be facing some serious water shortages.

.................................................................................................

-75-

**J. UBOIS** So as we started talking, we started converging and diverging on which problems were first and foremost facing us. One that struck me is that, as complexity becomes an important limb for policy, and SFI has done some amazing work on applying complexity theory to policy—from traffic patterns to city design to climate—it's going to be very hard to maintain the integrity of the discussions here as they migrate out and away and into policy circles, where for many policy advocates the problem is some crisis in legitimacy. What better way to get legitimacy than to say that the policies you come up with are based in the latest scientific thinking and the latest complexity theory from a place like SFI? That makes your policy prescriptions that much more credible, even if it doesn't necessarily make them that much better if you haven't applied the theory correctly. So I think in my nightmare list is this idea that in policy discussions the science done at the Santa Fe Institute devolves into something more like *scientism*. At least unless those of us here can get better at explaining what legitimate complexity theory is and does and what its limits are.

**L. SHEEHAN** Exactly, and I want to talk about some of the challenges that we're facing with our foundations. You know the rate of extinction is a thousand times the average rate over Earth's span. We're facing major climatic shifts that may have already pushed us into the sixth mass extinction. This sounds very alarmist, but unfortunately it's not just our foundations that are saying this. Scientists all over the world are saying this, and what we're learning from the work done at the Santa Fe Institute about complexity is really informative to law. I think that the law can evolve to reflect complexity in significant ways, and I'm really grateful for the opportunity to talk about this at the InterPlanetary Festival today, and to think about it in the context of space law.

As Jeff was just saying, the grounding or the scientific basis for law really illustrates its limitations. The Outer Space Treaty, which over one hundred nations have ratified and are generally operating by, was adopted in 1967. And, dating myself again, I remember watching the Moon landing shortly after that and being blown away by the fact that we could be out in the stars and that we may be masters of the Universe in a real sense. But complexity science is showing us the limitations of a design like that. That it's not just us pulling the strings—we're part of a web, which is great, but we need to be able to recognize and embrace that understanding in our law and policy.

**J. UBOIS** You had some good points about space law earlier, when we were talking about development in the legal community. Would you want to expand on that?

**L. SHEEHAN** Well, we know that we're interconnected. We know that we're interconnected with species, with ecosystems, with each other, with a larger Universe, and the Outer Space Treaty is very "cause and effect." So what does a system look like? What does a legal and economic system look like when it actually reflects the complexity, the complex adaptive system, within which we live? And really, I think that there are three approaches that we'd like to put forward today to talk about what we might think of as a potential legal and economic operating system that's consistent with science.

**J. UBOIS** Well, tell us about the Wild West. This is the first option.

**L. SHEEHAN** *Laughs.* Yeah, so we often hear that space is the new frontier, like the Wild West, and that we're going to go out there and that we're going to do good. We're going to address the challenges of today like the resource issues that we start to face, the limitations of what we have here on Earth, and we're going to solve that by going into space. It's exciting. But at the same time, it's fairly limited. The idea is that we are going to use the existing neoclassical economic models that were developed for the industrial revolution that pushed many of the advancements that we see today, and we're going to successfully apply that to exploration in space and all will be well. But I think that, as this InterPlanetary forum that we're all participating in today is telling us, we need to recognize the challenges that we have here on Earth, as well as in space together, and learn from both.

-77-

In thinking about this, Santa Fe Institute's Geoffrey West put a lot of thought into where we're going under our existing economic system, and what he says in his book, *Scale*, is that our current economic system is driving what he calls "superexponential growth," which will require infinite energy and resources within a finite amount of time. That's a problem! And we often think that we can technologically innovate our way out of it; we have a lot of faith in our technical abilities. But what he says in that book is that we can keep pushing that reckoning day outward, but eventually the time period for innovations is going to get smaller and smaller. We're going to have to innovate faster and faster to keep up with the superexponential growth limitations, and eventually we'll fail to do so.

**J. UBOIS** When you look at your peers in the legal community and the meetings that they're having around space law, a lot of it is focused on liability, and money, and the market considerations, and there is that idea that getting to space is something that can be supported through market mechanisms. But I wondered if we could maybe get in to the division between market-based government and philanthropically supported space travel.

**L. SHEEHAN** Well, sure, but why don't I turn that back to you? I don't want to be the one doing all the talking. I'll let you take that one. It's a great point.

### WIND BLOWS FURIOUSLY

**J. UBOIS** Wow, it's liftoff! So we were trading notes earlier about philanthropy, and one of the things that's true about it is that it has an incredibly broad diameter, and an incredibly wide range of activities that the eighty thousand foundations in the US fund, and that includes some space travel. So earlier this year there was a $20 million XPRIZE for a lunar landing that went unclaimed. But there was also a series of public comments by someone who's likely to become one of the largest philanthropists in the US, and that's Jeff Bezos, who said, "The only way I can see to deploy this much financial resource is by converting my Amazon winnings into space travel." That's basically it. Blue Origin is expensive enough to be able to use that fortune. Well, that was preceded about six months prior by a request for ideas. What are the things that one might do if, hypothetically, one wanted to move into philanthropy? There was a long series of suggestions, and now six months on we're back at all space travel all the time. So, anyway, I thought that was—

**L. SHEEHAN** No, that's a great point and you know, getting back to the three options for a legal system, the Outer Space Treaty specifically prohibits national appropriation of space entities. So the US can't go claim the Moon, the USSR can't go—I'm sorry, Russia can't go—and claim Mars. These privatizations of space are rejected, but, at the same time, with the type of pushback that we are starting to see, with technological need and interest in advances, we're starting to see the law change. So in 2015 the

US Congress and President Obama signed a space act that allows for private companies to go into space and claim property rights in space entities, which sort of opens the door for these types of mining activities that we've been hearing about. A lot of nations did not like this. It's very controversial because they felt that it was in conflict with the Outer Space Treaty.

There's been a lot of pushback and a lot of debate about what the law should be going forward. But in general this approach, this neo-classical/neoliberal economic approach that's driving superexponential growth, has become a big question mark. Is it the path that we want to be on, or is there another approach? There are two other approaches, one of which I know you've got a lot of experience with. The second approach is a commons law, a commons-based approach. Antarctica is a great example of this because nations got together and agreed to manage their use of Antarctica together, and to use it primarily for scientific purposes.

**J. UBOIS** I think this notion of property in space is going to, on the one hand, inspire a lot of people to invest in rockets and, on the other hand, probably give people headaches if they don't get there first. One of the other scenarios we talked about was the notion of a commons-based governance.

**L. SHEEHAN** That's right.

**J. UBOIS** So Antarctica is one example, and intellectual property has been another example where foundations in philanthropy have put a lot of money into promoting the world of open publishing and open intellectual property, as well as scientific publishing, but not too much for property. There have been some guesses about the value of ecological systems—can you somehow raise money against the future tourism revenues of the coral reef?—but that boundary between the market and the commons is uneasy and is probably going to be quite different in space. I wonder if you could speak to that.

**L. SHEEHAN** Sure. So the commons tries to push back a bit on this idea of ecological systems. Like waterways, for example, are often thought of in a commons-based approach: we should share waterways for all of us to use. We shouldn't have private rights to water. And yet we do have private rights to water. In California, where I do a lot of my work, we're seeing rivers run dry and aquifers being drained because of privatization of water rights. So we think a commons-based approach where we share the water exists in a different legal system, but both of those still suffer from a lack of attention to complexity. Both of those assume that natural systems—and, by extension, space systems—are objects, that they are set up to feed the economic system, that their use is property-based use, and that as we use them our economic GDP will grow and we should be all better off.

But again, circling back to what Geoffrey West has observed, this rate of exponential growth is perhaps slowed a little bit under a commons-based approach, but nature and the natural world is still an object. We still focus on managing for human desires as opposed to managing for what the system needs. And so what we've been doing is thinking about an alternative approach, a third approach that is more grounded in complexity, that recognizes more the interrelationships and the interdependencies that we all have with the world around us, our socioeconomic systems, and by extension, of course, the Universe.

**J. UBOIS** Another thing about this kind of approach we were discussing—

**L. SHEEHAN** The commons-based approach?

**J. UBOIS** Yes, the commons-based approach, and this idea of rights for those nonpresent, is this time dimension. And we had been thinking about what a long time horizon looks like for the development of law and policy systems. Some of the best philanthropy that has been done has been anticipatory; it has looked ahead, sometimes successfully, often not. But some of the older ethical systems around intergenerational equity may be applicable here in the commons-based discussions and in the space world that we've been talking about.

**L. SHEEHAN** Yeah, absolutely. You know, indigenous customary law generally focuses on both ancestors and on what is owed to the ancestors, the people who came before us, but then also looking ahead to the proverbial seven generations. How are we looking ahead? And the existing economic systems that we operate under are very present. They're very now. Even Adam Smith, who ostensibly invented the particular economic system that we live in, lived in a small community and worked in these community-based systems where the impacts of what you did were felt. So there was much more complexity in real time.

-81-

···································································

# We still focus on managing for human desires as opposed to managing for what the system needs.

···································································

Earlier today, somebody said that one of our problems is that we're too dissociated from the natural world. We're too dissociated from Earth. We don't feel the impacts of our actions. And when Adam Smith was writing, he said one of the ways that you could predict a state or a nation that's going the fastest to ruin is how fast it's generating and how large the profits it's generating are. And he actually lauded as wise and virtuous the person who puts public interest before their own private interest. This is not the label on Wall Street or on the existing economic system. It's more about recognizing the fact that we do live in a complex adaptive system and our actions do reverberate, and that the actions of others impact us.

And so, getting to the third option, moving from this sort of Wild West approach or the alternative commons-based approach, the approach that I think is most consistent with the idea that we live in a complex adaptive system is rights-based. This approach recognizes the inherent rights of all beings, including humans, but also everything that we have evolved to exist with, to struggle with. Not necessarily to be there forever and not necessarily to thrive—nobody can guarantee that—but to have a chance, to have the right to have a chance.

And there is precedent for this type of extension of rights. The UN passed in 1948 the Universal Declaration of Human Rights. And, in drafting this, the drafting committee said that they were not giving rights to humans: rights are not given by a king or a president, rights exist because humans exist. And by logical extension, rights exist because the Earth exists, because species exist, because space exists, and that recognizing those rights changes our behavior. The law is a legal and social compact; it's an agreement with each other that tells us how to behave. And in recognizing the inherent rights of others, we can modulate our behavior to respect the health and the relationships in the system which benefits all of us.

**J. UBOIS** So that kind of universal approach to rights, I would imagine, would exist in tension with an urge to decentralize authority in some way. Could you riff a little on the tension between the universal and the local?

**L. SHEEHAN** Yeah, I think that's a great question, and it gets back to the 1970s thought, "Think globally. Act locally." I think now we're saying, "Think universally. Act locally." Think about our place in the Universe, and then try to exercise what you can to make the relationships in your local community better. Thomas Berry is a cultural historian, and he wrote in *The Universe Story* that we are a communion of subjects in the Universe, not a collection of objects. It's not [that] humans are subjects of a larger governance system; we're all part of the larger universal governance system, and we need to regulate our behavior accordingly. We can do that by recognizing that our inherent rights exist and that other inherent rights exist as well.

And bringing that down to the local level is one way to do that. In fact, nations around the world—Ecuador, India, New Zealand, Colombia—and dozens of communities across the United States, from Pittsburgh to Santa Monica, are enacting laws and changing constitutions and winning court cases that recognize the inherent rights of natural systems to exist and struggle and evolve.

-83-

..........................................................................

The approach that I think is most consistent with the idea that we live in a complex adaptive system is rights-based. This approach recognizes the inherent rights of all beings, including humans, but also everything that we have evolved to exist with, to struggle with.

..........................................................................

**J. UBOIS** Okay, thank you. Should we take some questions from out in the audience?

**L. SHEEHAN** Yeah, I'd be happy to open it up. We have five minutes? Great, what a good time for questions. On that provocative note that nature needs rights, who wants to jump in?

**AUDIENCE MEMBER** Do you have any comments on how that would work in George's economics?

**J. UBOIS** How it would work for economics? Is that what you said?

**AUDIENCE MEMBER** George's economics, Henry George.

**J. UBOIS** Oh, Henry George! Single tax, land tax.

**L. SHEEHAN** Oh, man, it's been a long time since I've read that, but I could answer more generally. Do you want to jump in and take the Henry George part, and then I can answer it more generally?

**J. UBOIS** I think we might need more context for that, but it's about sending taxation out to the entire Universe to develop it.

**L. SHEEHAN** Well, I think absolutely we need—and I welcome your thoughts, Jeff, but just to quickly respond—we absolutely need to shift economic theory, and fortunately there are ecological economists that are developing economic theory to be able to do this. And what they articulate is that our economic system is upside down. We treat the well-being of the economic system, our GDP, as the end goal. And we talk about sustainable development or a green economy as if there are environmental solutions, but in fact the noun *development economy* shows where our priorities are. What they suggest is we flip it upside down and recognize that we're a part of a larger system, a Universe, with Earth, humans, and economics as a tool. It's a tool to be able to help these complex relationships thrive. I mean, what George wrote about is the mechanism of how to be able to implement that, but the thought behind what the goal is—maximizing social and ecological well-being—is very different than the current economic goal, which is to maximize the well-being of the economic systems and some people who are benefiting from it.

**J. UBOIS** One thing I'd add: It's very interesting to see what is being resurrected now. We've run into some problems, or hit the limits of certain ideological points of view, and so I don't know, it's a very subjective sense on my part that there is a look back at other experiments that have been tried in the last hundred years that are becoming more vivid again now.

**L. SHEEHAN** Well, absolutely. And, again, the idea of ecology as a sort of concept didn't come into the vernacular in a really significant way into the scientific community until about the early '70s ,which is when our environmental laws were passed: the Clean

Air Act, the Clean Water Act, etc. They're very reductionist, very separate; they say we can manage the environment for our own well-being. What the rights of nature approach offers is that we are more of a complex web, and that if we're trying to achieve healthy relationships we need to look at all the pieces and how they interact together, which is an incredibly exciting opportunity for science. It's how science is pushing forward in a complex way across disciplines. I see a hand up over there.

**J. UBOIS** Oh, yeah. Yay! Hi, Cory.

**CORY DOCTOROW** Hi. I really liked Jeff's interest in decentralization, but I think you dodged it a little. I think you dodged the decentralization question a little.

**L. SHEEHAN** Oh, I'm sorry. I forgot to answer it; I didn't dodge it.

**C. DOCTOROW** I think it's like, "Well, we like centralization when the EPA is run by someone who used to work at Greenpeace, and we don't like centralization when the EPA is run by a climate denier." So how do you resolve that tension?

**L. SHEEHAN** Yes, I have an answer. Sorry, I just forgot. I didn't dodge it! The idea of decentralization is twofold. One, it's to recognize those networks from which we've dissociated ourselves, and knowing where our food comes from, knowing where our energy and water come from, and being able to implement that locally. How do we treat our waste: Where does our waste go? That's something that we've put off onto others, but we need to reclaim that. It's really an honor to be able to live in a way in your community where you feel that you recognize and value those relationships.

So I think that there's a lot of movement forward, and that's what you were saying earlier. There are a lot of things that we've been starting that would allow us to be able to do that. And, from a utilitarian perspective, we're going to need to start doing that because climate change is starting to shift where we get things like rainfall. In California, we have a very centralized water system. We rely heavily on Sierra snowpack to provide our water. Well, that's not going to be there in thirty to fifty years in the way that we've come to depend on it, so we need to start to recognize local water supply. That's why Santa Monica passed its Rights of Nature Law to recognize the rights of the aquifer to be healthy, which forces them to take a different approach to how they manage that water locally. And they estimate that they're going to be self-sustaining in their local water supply in Southern California by 2020.

**J. UBOIS** That's interesting. Oh, is there one more? In the back? I thought I saw a hand up.

**AUDIENCE MEMBER** So if this was all decentralized and Santa Monica is taking care of its local water supplies, there are other communities which are using the same water supplies and that scales all the way up to countrywide scales like the United States. So you're still going to have the market forces that are traditional. What will it take to turn the larger scale into a sustainable economy?

**L. SHEEHAN** Did you want to take that?

**J. UBOIS** I don't know; I think replicable models are one answer. You have a lot of experimentation at the local level, and some of the projects that work well may win out. I don't think that's a great answer.

**L. SHEEHAN** I've done a fair bit of work on rights of waterways in California, and we have water rights in California for human use, and water has to be diverted and used in order to have a water right, but there are no rights in the waterways so they're getting drained dry. So doing pilot efforts on what a water flow right looks like, and how to implement that, that experimentation allows us to push back on these market forces which are using water for profit. I know that foundations are supporting this type of work in various states to some degree of success. But now that clearly is a problem because people need water to drink. Trying to build in this idea of a rights-based movement allows us to partner with social justice activists, with human rights activists, with people who are pushing for safe drinking water for everybody, because it's incredibly important. It's a right recognized by the United Nations.

But like any movement for rights, it doesn't happen overnight. You just have to keep pushing forward and putting forward models that you value. It took a hundred years for women to get the right to vote and that's pretty straightforward, right? Vote, don't vote. This is much harder, but if we put our minds to it and we offer examples and ways forward, I'm pretty optimistic of success.

**J. UBOIS** That may be our last one. Thank you.

**L. SHEEHAN** Thank you. Thank you.

APPLAUSE

**J. UBOIS** All right, well, we didn't get booed!

# INTRODUCTION:
# AUTONOMOUS ECOSYSTEMS

*The very act of creating and maintaining viable terrestrial ecosystems tests ecological understanding to the limit.*
—JOHN LAWTON

In a 1995 article in *Ecology*, eminent ecologist John Lawton defended the utility of what he called "Big Bottle" experiments—the use of carefully constructed, replicated, enclosed mini-ecosystems to study ecological dynamics. Micro- and mesocosms like Lawton's Ecotron—semiautonomous model ecosystems designed for research—have a long history in ecology. They have also been criticized for being too simple, too small, too short-lived, too unnatural, too specific, too closed, too far from equilibrium, too expensive. Lawton dispatches each complaint with alacrity and points out that "the Ecotron is a tool; like all tools it does some things well, some badly, and other things not at all."

Just a few years before that article, beyond the walls of academic ecology, another kind of "Mega Bottle" experiment took place— an attempt to create a sustainable, largely autonomous eco/social/ techno system in which humans, a patchwork of unique biomes, and a diversity of organisms could persist—a little spaceship Earth within spaceship Earth. From some perspectives, the two-year "closure" experiment isolating Biosphere 2 from Biosphere 1 (our own planet) was a failure—the oxygen disappeared into the concrete infrastructure and more had to be pumped in, some invasive species went wild, etc.—but in many ways it was a success, including what was learned from its obvious failures. Biospherian Mark Nelson points out some of those lessons in a 2018 article for *Geographical*: "the technosphere can be redesigned to support life without harming it; new roles for humans as atmospheric stewards and defenders of biodiversity; innovative biotechnologies to recycle wastewater and purify air; high-yield regenerative agriculture without use of chemicals; the importance of preserving natural biomes and putting limits on farming conversion; methods of restoring damaged ecosystems; shared dependence on the biosphere overrides group tensions and subgroups."

The Earth, as far as we know it right now, is the only "autonomous ecosystem," and even it will disappear billions of years from now when the Sun's luminosity increases and our star enters its red giant phase.

For decades, humans have contemplated what it would take to survive and thrive on distant planets. This has been expressed through much science fiction reaching back at least to the nineteenth century. It is also reflected in audacious experiments like the 1991–1993 closure of Biosphere 2, which was inspired by earlier Russian efforts and ideas; work by federal agencies in the US and elsewhere, such as NASA's HI-SEAS (Hawaii Space Exploration Analog and Simulation) program, whose first four-month mission was in 2013; and even through the popularity of little closed-glass terraria and aquaria available for purchase online. Of course, the ultimate answer to that question lies in what it takes to survive and thrive on this planet. The Earth, as far as we know right now, is the *only* "autonomous ecosystem," and even it will disappear billions of years from now when the Sun's luminosity increases and our star enters its red giant phase.

In preparing for our "Autonomous Ecosystems" panel, which included Biospherian Mark Nelson, HI-SEAS crew member Kate Greene, and digital artificial life explorer David Stout, I posed many questions that we might broach during our conversation. While we could only discuss a few of these issues in real time, they are well worth chewing on further, as I invite us all to do:

- ⊗ Is there any such thing as an autonomous ecosystem?

- ⊗ What are the biggest challenges to sustaining or creating an autonomous ecosystem?

- ⊗ What are the minimal requirements for a self-sustaining, persistent, resilient ecosystem?

- ⊗ Is there a difference between the simplest possible autonomous ecosystem and one that provides the kinds of functions important to sustain diverse life, including humans?

- ⊗ What are the key functions an autonomous ecosystem needs to provide, whether here on Earth or on some other planet?

- ⊗ What have we learned from the ways in which natural, human-impacted, and human-created ecosystems crash or transition to undesirable states?

- ⊗ Are we capable now of creating autonomous ecosystems? At what scale? Will we ever be able to? At a planetary scale?

- ⊗ Are there general signs of the impact of intelligent species on ecosystems or biospheres?

- ⊗ Can a nonliving planet be turned into a living one? How? Can a living planet be turned into a nonliving one? How?

- ⊗ What are ethical and practical issues around restoring, managing, changing, and creating ecosystems, on Earth or elsewhere?

- ⊗ What can digital ecosystems and artificial life platforms teach us about autonomous ecosystems?

The Earth's biosphere is a paradigmatic complex adaptive system, and one that increasingly bears the imprint of disruptive human activities that increase the risk of rapid, unpredictable, and undesirable state shifts at local to global scales. Thinking about how life can (or can't!) survive and thrive in other contexts can help us to understand how to enhance the robustness and resilience of this living planet.

*—Jennifer Dunne*
*Professor & Vice President for Science*
*Santa Fe Institute* -91-

# AUTONOMOUS ECOSYSTEMS

*David Krakauer introduces the panel, moderated by Jennifer Dunne and featuring Kate Greene, Mark Nelson, and David Stout.*

**DAVID KRAKAUER** Okay, let's go. You guys come up here with me, come up onto the stage with me. How are you doing, man?

**AUDIENCE MEMBER (DOUG)** Really good!

**D. KRAKAUER** Good, okay. You come over here, Doug. Beth, come over here. Wonderful. Okay. Hello, everyone. I have two extraordinary festival-goers standing on either side of me and—

APPLAUSE

**D. KRAKAUER** Yes, give them a huge round of applause for the extraordinary oath they've taken. I'm going to count down from three. Three, two, one, activate! Oh, yes. *Laughs.*

*Audience member Doug turns to reveal that his bald pate is painted to resemble a Roswell Grey alien head. Festival volunteer Beth Kiyosaki, dressed in a metallic dress and a metal dome-shaped hat, presses a button that sends the Solar System and astronaut action figure on her hat spinning around her head.*

APPLAUSE

**D. KRAKAUER** Fantastic! Okay, with that, let me introduce the next panel of minds. This is a panel dealing with an issue of enormous consequence on this planet—and on every other planet—and that is, how do you build robust ecosystems that span whole planets? How do you develop a sensitivity and an appreciation for the fact that what happens in the Amazon has implications for what's happening in your backyard? We have an incredible group of people here. They are going to introduce themselves. So, without further ado, "Autonomous Ecosystems."

APPLAUSE

**JENNIFER DUNNE** Can everybody hear me?

**AUDIENCE** Yes!

**J. DUNNE** I'm Jennifer Dunne. I'm a professor and the vice president for science at the Santa Fe Institute. My expertise is in ecology and ecosystems, and in particular in ecological interactions—how species interact with each other, particularly through feeding interactions but also all other kinds of interactions, through space and deep time. And so obviously I have an interest in ecosystems of all kinds, including potentially autonomous ecosystems. And what that means, what that is or may not be, we'll talk about later. Now, I'd like my colleagues on stage to introduce themselves. Maybe we can start with Kate.

**KATE GREENE** Thanks. I'm Kate Greene. I am a science journalist and an essayist. I've had the privilege of being a member of the first crew of the HI- SEAS experiment. This is a simulated Mars mission. It takes place in Hawaii. I was living in a dome for four months with five other people, and we pretended to be astronauts on Mars. It was all for NASA; they were collecting data to help potentially build a better Mars mission in the future. So I'm really happy to be here because "autonomous ecosystems" is something that came up a lot in our conversations. We were looking at social systems, as well as technical systems, but we didn't do as much on the ecology side. I'm very interested in discussing the possibility for space systems design to catch up and to think more ecologically.

**J. DUNNE** Great! And you?

**MARK NELSON** Hi, I'm Mark Nelson. I've been working with the Institute of Ecotechnics for decades now, and we set our goal in the early '70s to work on the integration between technology, all

human economy, and ecosystems. We did this in a series of resto-
ration projects in challenging areas. So I've spent decades here in
New Mexico, our semi-desert project. I've spent time in Australia,
in a similarly eroded tropical savanna, and working with our
wonderful project in Puerto Rico. My institute was a prime consultant to Biosphere 2. Biosphere 2 took the stakes up from resto-
ration ecology to ask, can you create a new world *de novo*? It was
an awesome experience, and I'm still digesting it. I think it was a
landmark project, and we really want there to be more biospheres
on earth as a preparation for what we will do in space. What has
particularly interested me in reflecting on that experience was the
intense and very satisfying connection that all of the crew felt with
that living system. And, as I look around at our ecological problems,
I'm thinking this is really the starting place. Humans are too discon-
nected and we don't understand, and therefore we treat our global
life–support systems badly.

**J. DUNNE** Great, thanks. And, finally . . .

**DAVID STOUT** Hello, I'm David Stout. I'm a professor at the
University of North Texas, where I direct the IARTA program,
the Initiative for Advanced Research Technology in the Arts. For
thirty-five years I've been working with interdisciplinary student teams to create large-scale immersive worlds, using physical
means and traditional media—actually, largely here in Santa Fe
some years ago. My personal work is more in the area of virtual
world-building: the use of VR and immersive video projection
and sound to model the complexity of interactive components, at
times focused on ecological and food-web kinds of interactions.
So I'm very happy to be here.

**J. DUNNE** Nice. Okay, thank you all for joining us. I just want to
start things off with the big fundamental question: *Is there actually any such thing as an "autonomous ecosystem"?* And, also, kind
of along those lines, *what do you see as the biggest challenges to
sustaining or creating an autonomous ecosystem?* I'm going to pose
that question directly to you first, Mark, as you're someone who

actually lived for two years in something that was trying to be an autonomous ecosystem.

**M. NELSON** Yeah. I will qualify that I think there's a problem with the term *autonomous*, because anything that humans build will have a lot of technology supporting it. If I can share an anecdote, we were really helped by Russians, who are the pioneers and are way ahead of what NASA has done to this day. And they told us, from their experience in closed ecological systems, that you can count on life to do what life has been doing for billions of years, and you can count on technology to break down—

LAUGHTER

**M. NELSON** So, I'd say, and I think from the Biosphere 2 experiment—which, given the unknowns we had to jump into, was remarkably successful—we're going to be able to build such systems much better. I think redundancy in technology and redundancy at the ecological level are two of the key components.

........................................................................

You can count on life to do what life has been doing for billions for years, and you can count on technology to break down.

........................................................................

**J. DUNNE** Just another question about Biosphere 2: Could you talk briefly about the different kinds of ecosystems, because there were several different ecosystems that were represented in Biosphere 2? If you could just give people an idea of some of those.

**M. NELSON** Yeah. Biosphere 2 had its space application, but we envisioned it as a laboratory for global ecology. We had a range of land syst—

MICROPHONE CUTS OUT

**J. DUNNE** Hold on. Your mic is cutting out, I think. Try it again.

**M. NELSON** Are you not hearing me? Okay. There we go. Can you hear me now? All right. Good illustration of technology breaking down—

LAUGHTER

**M. NELSON** You can count on it breaking down, the only question is, how badly and when? To summarize briefly, we had a range of land systems from wet tropical rainforest to savanna desert, mangrove, marsh, ocean, and coral reef. Then we had a farm and human habitat and all the technology needed to support that and to supply things that normally our biosphere does.

**J. DUNNE** Great. Well, we have two pod people with us.

LAUGHTER

**J. DUNNE** Here's our other pod person, and it was a very different flavor. Can you talk a bit about the NASA experiment?

**K. GREENE** Absolutely. It was very different. We tried to keep the ecosystems out, we were really focused on the social systems and the technological systems to support the people.

Now what that meant was, we had solar panels and generators for electricity, and our water was brought in and pumped out. So we were not a closed system at all. We were really interested in looking at what people do in isolation and how that might be useful for a Mars mission. That said, ecosystems certainly crept in, so one example is we had algae in our water supply. This really changed the way that we behaved.

Some people drank less water, which wasn't good. We used more energy to heat the water, to potentially sterilize it. You have these unexpected ecosystems, you can't get rid of them, even though we were trying to do that in a way. It really changed the way that we acted as a group and the technology that we used. So I think it's incredibly important to consider the interplay between all of these systems, the technological, the social, and the ecological.

But Mark had so many more complex things going on; he had all three of those systems. So I'm really interested in hearing where things were done for you guys and where you learned that resiliency needed to be the most important focus.

**J. DUNNE** Yes, but before we move on to that, I just want to give David a chance to, from the artistic digital perspective, talk about what an autonomous ecosystem is or is not.

**D. STOUT** I would look at it as a hermetic exercise. And within a digital system, where we want to model agent-based behavior, it's very much a closed system, but it's also a technology guaranteed to break down. In doing something like that, where lives are not necessarily at stake, you can make huge speculative leaps in terms of what you're modeling and the kind of data that you might want to mix together.

In its relationship to the actual real-world experience we're talking about here, art is able to give us, or it allows us to examine, the ways we represent ecosystems and how we think about them. The very term *environmental* or the term *ecological* as they entered into the mainstream vocabulary, I would say across culture during the '70s, very much underpins the way we think about a lot of things. It isn't always precise, and so within an art project we can take various kinds of dynamics like the kinds of things you're doing, Jen, and model them in various kinds of spheres.

**J. DUNNE** Can you actually talk a little bit about the "Hundred Monkey Garden," which is the piece I'm most familiar with?

**D. STOUT** Yeah. We did a project called the "Hundred Monkey Garden," which was essentially looking at a computer system as having a limit to growth that can only do certain kinds of things, and using that system resource to its hilt. We had two parent forms that could procreate, based on core data and separate data streams, roughly like a kind of exchange of genetic material. If they were successful, they could populate the screen with all kinds of flora and fauna. And, existing on a separate computer system, was a predator that, when it was successful, was eating their children—

LAUGHTER

**D. STOUT** It would grow more and more elaborate over time. What would happen is, if the parents were really successful, the computer, while it never actually crashed, it would slow down to such a speed that literally it was like one cycle every five minutes and nothing was happening.

LAUGHTER

**D. STOUT** So at moments it was very well tuned, and there was this balance of procreation and nutrient death, we'll call it, and things would run really smoothly. That would be an example.

**J. DUNNE** Can you speak briefly about what's on the screen behind us?

**D. STOUT** Oh, yes. The images you see here are from more recent works that I've been doing with Cory Metcalf and a few other team members. We're modeling various kinds of plausible worlds that are at times alienesque and at other times very much like Earthscapes. At the moment, we're about to launch a new project at the CURRENTS exhibition tomorrow that populates these environments with various kinds of bacterial or viral forms that can actually transport you from one place to another.

**J. DUNNE** Great, thank you. Now, getting back to redundancy/resiliency factors, a number of lessons were learned, I think, out of Biosphere 2. It would be fun for you to talk about a few of those, Mark, if you can. A big lesson was what was going on with oxygen.

**M. NELSON** The oxygen?

**J. DUNNE** Yeah. I think that's a great story to tell.

**M. NELSON** Yeah. It was. And there are a few things about oxygen that make it particularly kind of illustrative. One is that, you know, at project meetings for years, we'd go around the room and people would give their ten greatest nightmares—what problems they're worried about—and nobody had ever said the oxygen would disappear and it would take a lot of scientific sleuthing to find it. So it's curious, because in the popular press, somehow they forgot that Biosphere 2 is a pretty kickass, very audacious experiment. It wasn't perfect from the beginning, but the really beautiful thing is that, when Columbia University first got involved with Biosphere 2, with really careful scientific sleuthing, we could actually find out where that oxygen had gone. It reflected what is not that surprising in retrospect—there was an imbalance between photosynthesis and respiration. Of course, we started with a very baby biosphere, and some of the plants grew from five or six feet to thirty and forty feet. The calculations showed that, by five years later, the system would have stabilized in terms of oxygen.

**J. DUNNE** There's also another story that's an interaction story, between the physical structure, the concrete, and what was going on, biotically.

**M. NELSON** Yeah, and if you haven't heard these terms, these come from the Russians. They talk about a *biosphere* and a *technosphere*—and this was a reflection of what my institute was looking at—there was a technosphere in Biosphere 2, and it was pretty huge. Even for a three-acre miniworld, you need miles and miles of pipes and innumerable pumps and gadgets and "witchy-watchies." But we had to make our technosphere in such a way that it wouldn't pollute the biosphere. And that was a really interesting exercise, to get engineers to understand that we don't just want a really elegant, technical engineering solution. You have to remember that this is going to be in a virtually closed system, and it cannot damage the life. It was a beautiful illustration.

But it was the unsealed concrete inside Biosphere 2 that was absorbing the oxygen in the form of carbon dioxide molecules. And to me, if ecology means everything is related to everything else in our global biosphere, certainly in Biosphere 2 everything is connected, including the technosphere.

**J. DUNNE** Absolutely. I think this is a recurring theme. You mentioned it too, Kate, with the algae in the water. There were probably other things that happened with the NASA project, but these supposedly closed systems are not-so-closed systems. There are unintended consequences as we try to construct these things. It's hard for us to replicate, in a dynamic, persistent way, the stability of the system over time, and it requires a lot of energy and thought and action. You were in Biosphere 2. Of course the original biosphere is the Earth. The Earth is really the only autonomous ecosystem we know of at the planet level, but it's getting a ton of energy from the Sun that is fueling all of the biodiversity and interactions and things that are going on even in cultural development.

So it's not just about what the technology is that's built into it. It's not just about what species you introduce. It's not just about human practices. It's about all of these interactions. It relates to the kind of work I do, which focuses on interactions. Once you get these complex networks of organisms and technology and the physical environment, you drastically increase the chance of having unintended consequences. It's something really important to keep in mind.

**K. GREENE** Well, what you're talking about, Mark, the problem of concrete absorbing oxygen, it just makes me think about the assumptions that we make when we build these systems—what gets included and what gets left out—and what we find out about later is what we usually call "unintended consequences." But, actually, it's just a lack of making proper assumptions on the front end. And that's what you do with modeling. When you're modeling systems, you have to ask yourself these questions. What are the bits that you include? Then you see how those things interact and play out. So I'm interested in asking you, Jen, what sort of assumptions do you always include when you're modeling these complex ecosystems, and what might you add later as you ask different questions?

**J. DUNNE** Yeah. Any time you're doing any kind of scientific modeling, any particular model is going to be good for certain kinds of questions and not others. There are some very simple models I work with that look at how feeding interactions are structured in the context of a food web, and it is just giving you the architectures of what is going on between species. So it's good for certain kinds of things—for example, trying to understand the balance of generalists and specialists in systems. For example, I would say in any Biosphere 2, you'd want to set up, you'd have to be conscious of the fact that you'd need a mix of generalists and specialists. There is much more complex dynamical modeling that we do that includes a lot more kinds of assumptions. They make the model more complex, and harder to understand, but that also allows it to do more and to answer other kinds of questions. It's really context dependent. But it's critically important, because in the case of the NASA experiment, there was a much more limited scope of what they were trying to look at.

**K. GREENE** Right. In our case, we were just looking at two different food systems. We were asking the question, Might it make sense to bring along shelf-stable food that you could cook with once you landed on Mars? Like, could you bake a cake to celebrate landing or a birthday party? Could you have a pizza party or something like that, which is something you couldn't really do with the typical NASA add-water-and-heat food, which is what the ISS astronauts have. So we were looking at just-add-water-and-heat food and then the raw ingredients that you can use to be creative and make your own food. Those were the two systems that we were looking at, and that's actually one of the reasons why we didn't have a lot of plant life or an ecology that we were working with, because we wanted to keep that food system isolated. And we found out that we really like to eat meals together, we really like to celebrate with cake—

LAUGHTER

**K. GREENE** —so that was one of the things that came out of that experiment.

**J. DUNNE** Mark, can you talk a little bit about food systems in Biosphere 2?

**M. NELSON** Oh my god, they were so important! Absolutely. We didn't expect this to be the case. We got to be better farmers; we weren't the most brilliant farmers at first. There was a lot of cloudy weather, etc. Our doctor was really happy. We were on a low-calorie, high-nutrient diet; i.e., we were very hungry. We were hungry the entire two years, and it became super important, very quickly, that the cooks really make interesting meals. Even though we had forty-plus species, the cooks needed to make new presentations. And we also collectively decided that we were going to put a little bit aside, because then we could look forward to a holiday, a big feast day, and that became a really important part of the social knitting of the crew. A bad meal was just a disaster for crew morale!

**J. DUNNE** *Laughs.* And can you talk a little bit more about what you've learned from feeding-type interactions in your digital systems, David?

**D. STOUT** Well, coming out of an art project, as opposed to doing it from a scientific perspective, there is a certain naïveté involved. When you actually start to contemplate the complexities of nutrient systems and interactions, and then think about how you might model them within the limits of the technology, it becomes kind of mind-boggling. And so one has to get very inventive in terms of how you represent different kinds of data. That's of course one thing that we're grappling with, with the project that you and I are contemplating. I'll talk just a little bit about it, because it relates to this idea of pods and autonomy—that is, modeling multiple ecosystems that have different kinds of constraints, but then opening up wormholes between them at different locations and allowing one species or one nutrient source, or other elements that govern those individual pods, to interpenetrate each other.

It is very much an exploration of autonomous ecosystems, but it's about polluting them, or opening them up to other kinds of migration. Obviously, we are all, in a sense, autonomous ecosystems running around, but we're anything but autonomous in our social interactions, or given all of the viruses that we encounter, etc.

**J. DUNNE** Absolutely. And, in the context of the project you mentioned, a lot of it is looking at invasion dynamics—things spreading throughout networked systems and what the impacts are. There's a lot of work in ecology that studies patch dynamics and invasion, and some of these things can sound negative in a non-science sense, but actually they can be very important in terms of the resilience of the whole overarching system, of many systems linked together. That whole notion, and what Biosphere 2 was trying to achieve, of having different habitats and different ecosystems represented, is getting at or trying to look at the importance of that diversity, of different kinds of systems that are very deeply connected.

**M. NELSON** Yeah, and even when I said that we had rain forests—the ecological design was so sophisticated, because nobody knew, of the thousands of species that went in, would we lose ninety percent of them? Twenty percent? Sixty percent? But within each system like the rain forest, there were different habitats, and we wanted to maximize both the microbial levels and also different terrestrial levels. So there were six different ecological patches or zones within the rain forest alone.

## We like this space—we're going to conquer it.

**J. DUNNE** Yeah. Again, hierarchies. Layers. Interconnected systems.

**M. NELSON** And then we have the Biospherians intervening under the banner of biodiversity. So while you were talking about invasive species, I was thinking about morning glories, which not only threatened to swallow the rain forest but were poised to enter the savanna.

**J. DUNNE** Were morning glories brought in, or did they work their way in somehow? Through a seed bank or . . . ?

**M. NELSON** Well, I've gotten two different answers, but they were definitely designed for a very limited habitat, and they decided, "We like this space. We're going to conquer it."

**J. DUNNE** Yes, and we see that in ecosystems: many invasive species aren't successful, but it only takes a few to strangle the dynamics of a system and fundamentally change it.

**M. NELSON** That's another part of that *autonomous* term. If there's a system that involves humans, we want astronauts to be ingenious—I should have said that we became way better farmers than where we started. We grew a ton more food—there's that element, that ecology and humans are adapting. We also thought a hundred years from now, at the end of the Biosphere 2 experiment, the early technology will be totally outmoded. So you have to build in room for evolution at every level.

**J. DUNNE** Okay, so, to get back to fundamentals of this festival, are we capable of going to another planet and creating a sustainable biosphere? And, if so, what would it look like? We have two very different models here. Which one is more likely to be successful?

**K. GREENE** From what I've seen, I don't think that an autonomous ecosystem would be a part of a Mars mission. I think that space system designers will do everything in their power to keep the ecosystems out. You might see some gardens for pick-and-eat food, so maybe some salads that an astronaut might have on the way, but no one is going to be depending on crops, because ecosystems are unruly by their nature. We can model them, but we don't really know what's going to happen, and then when you have people intervening, it becomes even more complicated. I think the idea in going to Mars is to keep things as tidy as possible, and ecosystems are not very tidy.

I just think about the Martian ecosystem that an actual Martian habitat would have to contend with. You would want to have this thing that's impervious to the outside, but the dust on Mars has salts in it called perchlorates. So this would be a superfund site, you know, but if you wanted to put your house down, where we would be putting a Mars habitat, these things get inside that could really cause some problems. Ecosystems will always creep in, even when we don't want them to be there.

**J. DUNNE** What's the most optimistic promise of our lessons from Biosphere 2?

**M. NELSON** Let me just say, Biosphere 2 is a vision of long-term evolution in space where we could make it work using local resources. Clearly, as Kate was saying, we're going to start with probably hardly any food production. But when you start to have people living in space, in orbiting space stations or on the surface of planets or moons, the cost of resupply becomes extremely high. So I'm really not sure. I'm not a big fan of hydroponics, whether you'd even say a hydroponic system is an ecosystem. The Chinese, based in Beijing, recently concluded a one-year operation with two or three crew members who produced all of their food, grew meal-worms—the Asians are much more happy with eating insects than most Westerners—and that was quite successful. Space life support is going to start very limited, with robust backup. But our contention is that eventually, psychologically, humans can't live in a little hydroponic chamber. Psychologically, we come in the package of a biosphere. And, if I picture living in Biosphere 2 and it was just hydroponics, it would not have been a happy two years.

**J. DUNNE** Right, and I would also say that art has a role to play. I mean the importance of culture, artistic expression, music . . .

**D. STOUT** Well, in terms of thinking about this, the original impetus we had as hunter-gatherers we've depicted in various ways. Now that we have an understanding of some of these complexities, art has to sort of match the scientific inquiry. But, in terms of the question you just asked, my interest as a communicator is more in populating space with communication signals and other forms of art that are sent out into the cosmos. I'm sure there's another panel that's dealing with alien contact—

**J. DUNNE** Yes, Pete Worden will be talking about that later.

**D. STOUT** But obviously when you're talking about speculative ecosystems, you really cannot ignore that.

**J. DUNNE** Absolutely. I think on that note, we've run out of time, but I'd just like to say that David will be at CURRENTS; he has a piece at CURRENTS. And Mark, you have a book here?

**M. NELSON** Yes. Synergetic Press has a booth over here, booth five. It's got great books.

**J. DUNNE** And, Kate, I don't know if you have anything that you want to promote?

**K. GREENE** No, no. But I'll be here all day!

**J. DUNNE** Yes, we're going to be around. If you have any questions about any of this, please come talk to us. We're happy to talk to you.

# INTRODUCTION:
# TIME DESIGN

"Time is on my side," or so sing the Rolling Stones. But is time on the side of us Earthlings? Almost without exception, we think that having extra time is being on the better side of time. This often means getting tasks done faster and more efficiently, but it can also mean just living longer.

When it comes to interplanetary colonization, are our planet and civilization on the better side of time? What are the timescales or clocks involved in even asking such a question?

For interplanetary existence there is a set of innovation clocks that time how long it takes to develop the necessary inventions—long-distance galactic travel, breathable environments, astronomical agriculture, space sustainability, and manageable levels of gravity and sociality that create the ties that bind. Equally important is another set of clocks, which we'll refer to as catastrophe clocks, that count down until we're forced off the Earth by extreme conditions and dangers imposed by climate change, exhaustion of resources, nuclear winter, or the rivers running dry. Naturally, what matters is whether we can create the needed innovations before the catastrophe clocks count down to zero.

Recent work at the Santa Fe Institute helps shed light on these questions. In work by Geoffrey West, Luís Bettencourt, and collaborators, we see the catastrophe clocks accelerating in speed. The innovation clocks have been able to keep pace—so far. But how long can this last? What can we do to accelerate the innovation clocks and to slow down the catastrophe clocks? If we migrate to other planets, will that enable us to reset our catastrophe clocks while leaving our quickly ticking innovation clocks intact? Or, if we are colonized by creatures from other planets, will that accelerate our count-down clocks, our innovation clocks, or both? Can we consciously uncouple these two sets of clocks or other types of clocks in a way that will benefit our survival and our humanity?

These questions seem straightforward to define—if not answer—while sitting here on Earth. But even posing these questions becomes subtler when we take seriously the full implications of interplanetary existence. What happens when we realize there is no universal clock? That is, across the Universe, across planets, across species, across cities, and while traveling at different speeds, clocks will all tick at very different rates. Whether we start from the most fundamental definitions of time—sequential events and increasing entropy—or the most applied—changes in day versus night and seasons—we realize the uncoupling of interplanetary clocks is inescapable. Yet to coordinate efforts on Earth with those on other planets, we must find a way to couple some of these clocks simply to allow for basic biological needs that will let us eat, sleep, and live.

## What happens when we realize there is no universal clock?

These concerns are exacerbated when we consider even more of the strangeness of time. For instance, what if time is not just linear but instead cyclic or multidimensional, as conjectured by several cultures and allowed by the principles of theoretical physics? Although making sense of cause and effect and the sequences of events might start to boggle our brains, such possibilities might be literal lifesavers when it comes to finding ways to avoid the countdown of catastrophe clocks. This could either be by circling back to earlier times, taking a detour around dangerous times, or funneling energy off from other places in time and space in the larger Universe.

And what about advances in artificial intelligence, robotics, machine learning, and cyber–physical interfaces? How much will these be able to extend our lives and change our individual, biological, and societal clocks? Moreover, how much can these accelerate innovation clocks due to the many orders-of-magnitude increase in computational speed of today's fastest computers in comparison to our human brains? And how much more will this be in a few years, assuming the continuation of Moore's Law—one of the most famous innovation clocks—that shows computational speed and power are increasing exponentially in time?

Some of these innovations seem within our grasp, and some may seem more far-fetched and might only happen after we have already migrated to other planets. Or maybe they will never happen. Or maybe they will be brought to us by visitors from other planets. One of the most mysterious and frustrating things—yet also one of the most hopeful things—about our Universe is that we can't always know or predict the future because we are stuck in the eternal present, we remember the past, and we are mostly blind to the future. Time topples all (entropy) but time can also be restorative—"this too shall pass"—and bring back light (cyclical). Only time will tell where we end up, both in terms of survival and on which planet!

These are the small and large thoughts that were on the minds of the panelists—Stefani Crabtree, Martine Rothblatt, and myself—when we sat down to talk about time at the InterPlanetary Festival. Our time was limited—half an hour—but all of the topics above were raised and pondered.

Placing or positing more definite rates on each of the innovation and catastrophe clocks would be a fun and much-needed exercise for the future that would help inform how hopeful we can be. Perhaps this is a topic for the next SFI InterPlanetary Festival. We hope you all keep tuning in for many years to come and follow us with whatever clock and to whichever planet we end up on. If friendly interplanetary visitors drop by—like Zaphod Beeblebrox picking up Trillian in *Hitchhiker's Guide*—life might be closer to a Beatles lyric: "Working for peanuts is all very fine, but I can show you a better time."

Hold on to your helmets, and prepare for an exciting journey. May time be on the side of us all!

-111-

—*Van Savage*
*Professor, UCLA, Santa Fe Institute*

# PANEL: TIME DESIGN

*D.A. Wallach introduces the panel, moderated by Van Savage and featuring Martine Rothblatt and Stefani Crabtree.*

**D.A. WALLACH** Hello, everyone. This is my first panel introduction of the day. We're going to be getting to know each other very well over the next few hours, as I will be taking over emcee responsibilities from David, who we should all congratulate on his terrific job so far.

APPLAUSE

*D.A. looks out into the audience, where a number of people have brought their own chairs and are gathering closer to the stage.*

**D.A. WALLACH** And as I heard Stefani expressing, I have to give serious props to the folks sitting in the front here. This is courageous and noble, what you're doing. So much respect!

**STEFANI CRABTREE** You guys need to calm down!

**D.A. WALLACH** This panel, in addition to taking us to *eleven*, is about to go extraordinarily deep because we're about to get into time, which, in our new interplanetary view, has to be a central concept for us to consider. What is time? How do we think about the scale of time when we're dealing with the entire Universe? When we're not just living on our planet but potentially other planets? How do we think about the scale of our own lives? Our temporal existence? How do we think about new technologies? How do we manage our life spans? What should our goals be in this respect? How do we think about generations?

And with us today we have three amazing folks. And Michael here, Michael Garfield, is going to be doing generative scribing during the discussion. That will represent the artistic component of this panel, but let me first introduce the panel.

Martine Rothblatt, with whom I have thankfully just spent some time, is a magical mystical person, whose written bio, which I won't read in full, begins with the best sentence I've ever read in a bio: "Martine cultivates invisible parts of reality," which is extra-ordinarily trippy.

APPLAUSE

**D.A. WALLACH** Her career began with SiriusXM, cultivating the electromagnetic spectrum, and since then she has done a number of remarkable things. But, and you can learn about this at the expo booth over there, United Therapeutics has become her vocation. She's developing therapeutic treatments for people with diseases that have gone ignored for too long, and thankfully, when addressed, represent both great humanitarian opportunities and great commercial opportunities.

Stefani Crabtree is at the Human Environment Dynamics Laboratory at Pennsylvania State University. Her work looks to the past to give grounding to the way we think about the future, using all sorts of dynamic simulations, agent-based modeling, and other fancy techniques that I, myself, don't understand. Stefani is trying to understand whether the patterns of history—and correct me if I'm wrong once you start talking—can give us any sense of what's to come and can allow us to make predictions about the processes in which we're currently a part.

**S. CRABTREE** Yeah.

**D.A. WALLACH** Van Savage, arguably the best name on the panel, if not in the Universe, is at UCLA but has deep roots at the Santa Fe Institute. Van is a theoretical physicist studying quantum gravity, but, ultimately, he has branched into areas as exotic as the biology of circulation in humans, life spans, and what biological time means, correct?

**VAN SAVAGE** Yes, that's right.

**D.A. WALLACH** So I've just given you a small sense of what's to come and with that, please welcome our panel.

APPLAUSE

**D.A. WALLACH** One last note: After the panel, Martine is going to be signing copies of her book, *Virtually Human: The Promise—and the Peril—of Digital Immortality*, right over there. So get a copy and get a signature!

APPLAUSE

**V. SAVAGE** Hi, can everyone hear me?

**AUDIENCE** Yeah!

**V. SAVAGE** As D.A. said, I started off in physics and then switched to biology, so I'm going to talk about time from a very broad perspective: what time is, how we measure it, how that translates into physics and biology, and how those clocks coordinate. I think that will lead into questions about civilizations and cities that the other panelists can better speak to. I'll talk a little bit, and then I'll turn it over to the panelists for the rest, and then we'll take questions at the end, if there is time.

My background was originally in theoretical physics doing charge parity time symmetries and time reversal, so a lot about what time means. The question we can start off with is, how do we define time in physics? One way to do this is just by the sequence of events. For instance, I had to walk up on this stage before I could talk to you. Or, if I hit another car at the intersection, it had to get there before I did. So certain events have to happen in order, and that defines time.

We're also going to hear more about the idea that, according to the laws of physics, disorder increases in time, so things become more disordered and random over time. So if you watch disorder increasing, that's also a marker of time.

Of course, that's not something that we can easily measure. If we want to talk to each other about where we should meet at three o'clock, or where to meet tonight at ten o'clock to have a drink—when we talk about time in our everyday lives—we usually mean clocks. And how we've developed our definitions of time, how we've been thinking about time, started with very basic astronomical and seasonal facts. A year is how long it takes to go around the Sun. A month is related to the phases of the Moon. A day is related to one revolution of the Earth.

In time, we got more and more sophisticated and began to think about things in different ways. For instance, now the current definition of a second is how many decays there are of a cesium atom, or the number of transitions of energy levels in a cesium atom. And one thing we'll come back to is the fact that coordinating clocks is a nontrivial thing to do. It's very important for getting things right. That's something we see in our everyday life.

When train systems were built, we had to coordinate clocks for different trains to make sure that the trains arrived and left when they were supposed to, so that we could communicate with other cities. Now, in terms of time, if we have an application for something due at five p.m., my computer time better be the same as your computer time or your phone time. For financial transactions, I can make money if my clock goes at a different rate than someone else's. If the clocks are mismatched and things are set at different rates, I can make money from that. There are actually financial companies that pay to have huge lines built to increase the speeds of their transactions, to take advantage of clock mismatches.

So that's a case where we all sort of agree on our clock. We're all on Earth, and we all agree about what time is and what our clock is, but in order to tie this back to interplanetary thinking—in terms of colonization or communication or travel to the planets—I'm going to argue that these problems become much more complex and much more sophisticated. For example, if we think about going, in the most trivial sense, to another planet, our definition of a day or a year might change, right? It may take us longer to get around the

Sun, or there could be two suns, or we could rotate much faster or much slower.

Now, you might say "Well, okay, sure, but I can still put a clock there and have a clock here on Earth, and if the clocks are set by the same time standard, they'll still measure the same time. If I talk to them on the other planet, I can keep the same time." But even that's not really true, due to Einstein's theory of relativity. His whole point is that how fast the clock ticks, or how fast the second hand ticks, depends on how fast things travel. The faster you're going, the faster the speed, the slower that clock ticks, actually.

And then, if you go as fast as light, the clock stops ticking altogether. Light, in terms of Einstein's relativity, doesn't experience time at all. So I've heard this question, "Does time exist?" My answer often is that it does, because light doesn't experience it. So it knows what the absence of time is. That's one way to think about it.

So now I'm going to speak about biological time. We each, even if we don't have a watch, carry about lots of little clocks inside of us that tell us when we should sleep, how fast we grow from birth to adulthood, and how fast we age. We have clocks going around our bodies all the time, but we're just not conscious of them.

If you go to another planet, because of the daylight cycle or because of the temperature, for different reasons, those body clocks may be reset. If you think about a complex ecosystem with fish or birds and mammals or lizards and reptiles, some of those things—the fish, the lizards, the reptiles—if you change temperature, their clocks change essentially. They age faster, they grow faster, they do every-thing faster at higher temperatures, whereas we don't. So we're sort of decoupling clocks. We're changing clocks of some organisms, but not of others.

It's not even as simple as one single biological clock. We have many biological clocks. And even for mammals, we have a constant tem-perature, but a mouse has a very different clock than an elephant. Its heart beats much faster, it ages much faster, it gets cancer much faster. In thinking about the interplay between all these species we need to coexist, our predator–prey interactions, and what eats what and how that may change on a different planet with different temperature, different daylight cycle, is really a complicated endeavor. Like, how do we couple these biological clocks? How do we couple those biological clocks with physical clocks, or trains?

### TRAIN PASSES

That also leads into not just physical clocks and biological clocks but things like clocks for civilization or cities, and thinking about city clocks. Some work done at the Santa Fe Institute by Geoffrey West and Luís Bettencourt and several other people shows that the larger the city, the faster the clock ticks, as well. The people tend to walk faster in larger cities, do things at a faster rate in larger cities. Feel more time pressure in larger cities. Even our clock across the country can vary depending on what city we're in. Not the clock we measure on our phone but the clock in which we think we need to do a series of events.

........................................................................................

In space, we can walk forward, backward, left, right, up, and down. With time, we tend to be able to only go forward. We can't go backward. We often wish we could, but we can't do it.

........................................................................................

That relates to not just us getting around every day but also timescales for civilization advances, maybe travel to other planets. I guess I always thought I was immune to those city clocks. I would sort of stay up all night and sleep all day. But now that I have a child and a wife and a job that I have to go to regularly, once you couple—that's the whole thing about coupling, is once you have to integrate with lots of other people and lots of other schedules, you can't get away from that. Those other clocks certainly matter to you.

Now, I promised David I'd talk a little bit about quantum physics. And that will tie into things very well, which has to do with our perception of time. What I've been talking about so far is really coupling timescales or clocks in ways that allow us to coexist in species and be able to eat and survive, have a daylight cycle that works for us. But, also, to be able to think about these questions and how we perceive time matters.

-119-

We'll hear more about this from the other panelists as well. But even in physics, there are very different ideas for how to perceive time, even in quantum physics. One of the ideas out there is this idea of a multiverse. That, in quantum mechanics, you have a choice at every moment—say, one electron could be spinning up or one could be spinning down—and that leads us to all the possible things that could happen in the Universe and all the possible branches and choices we make.

Actually, every time that happens, we branch into different universes. We actually have this ever-expanding branching of multiple universes. And if there was a way to travel back and forth between these universes, it would change our perception of time and change our perception of space because the sequence of events that would happen in each universe would be very different.

Also, one final thought before I turn it over is that we often hear about one of the things that separates time from space. In space, we can walk forward, backward, left, right, up and down. With time, we tend to be able to only go forward. We can't go backward. We often wish we could, but we can't do it.

But the other question that I always ask is, not only why do we go forward, but why can't we have multiple time dimensions? Why do we only go forward and backward? Why can't we go left and right and up and down? Why can't we circle around in time in different ways? And if physics is not disallowed, it creates all kinds of complications, but it's not impossible. It could happen to allow extra time dimensions, and that adds even more complications when we think about how we couple time clocks, how that might work as we travel to other planets, or talk to other civilizations. I think we'll hear about those ideas from some of the panelists as well.

I think I've talked enough, at least to start. I guess I'll come back later and take questions. But I think Stefani should talk now.

**S. CRABTREE** Great.

**V. SAVAGE** Doctor Stefani.

APPLAUSE

**S. CRABTREE** Hi everyone! Time is an interesting question for archaeologists. Because we spend so much time thinking about the past, and I think that, in all honesty, archaeology can save the world. We're all dealing with some pretty big challenges now. But we can look back into the deep time and see other civilizations that have lived through similar challenges. That's, I think, one of the reasons why bringing in archaeology to talk about these questions of time, to talk about interplanetary health, is important because we can look at the trajectories of civilizations, the ways they experience their lifetimes, as individuals and as societies, and look at the different challenges that they face and the ways that they confronted them.

Some of the ways that I approach looking at the archaeological record are through computer modeling. I also dig in the dirt. But I spend a fair amount of time building simulations, network simulations, and *SimCity*-style simulations, agent-based models, where we can build these worlds that experience time and run them hundreds of thousands of times to look at what happened in these societies. It's an interesting way to be able to have a time machine, essentially. To build these simulations based on things that we know about archaeological societies.

Running back around to this idea of time and how societies experience time: We now think about time in a linear way, right? My birthday is the end of June, and so I'm going to mark the fact that I will be a year older. We are experiencing the year 2018 right now. Next year will be 2019. We look at it in a linear manner. But a lot of societies, we think as archaeologists, experience more of a cyclical style of time, where you're dependent on the seasons. Where, instead of looking at the clock moving forward, and calendar moving forward, we are looking forward to another spring and we're resetting the clock every time. Or if we look at Mayan civilizations, we know that they had a four-hundred-year-ish cycle for their calendar. And so there are very different ways in which we can experience time, as Van brought up.

–121–

That we need to, as a society, understand our own internal time, understand the ways in which we all experience time and relate it to each other, the ways it relates to how society is progressing, and how we're interacting with the Earth, and with other things we're interacting with. And these will be challenging questions we're going to face if and when we do end up going out into the stars and establishing colonies out there.

*Technological difficulties arise with the microphones, and the panelists work together to resolve them.*

**S. CRABTREE** See, now we're experiencing the time delay of microphones!

**V. SAVAGE** Yeah.

**MARTINE ROTHBLATT** Hello? How's that?

APPLAUSE

**M. ROTHBLATT** Okay, thank you, Stefani. Now I can clap, too. And just in the nick of time. So I'd like to talk about time a little bit, at the same time as mentioning three people who have influenced my thinking on this subject: Erwin Schrödinger, Ray Kurzweil, and Gerard K. O'Neill; and talk about time a little bit from the standpoint of physics, a little bit from the standpoint of human perception, and a bit from what I do in my day job, which is technology development.

So I think of time physically as something which, as Van mentioned, represents an increase in entropy. People say that ultimately if the Universe completely ran down according to the second law of thermodynamics, we'd be in a state of maximum entropy, and we would reach kind of an end of time where nothing changed anymore.

It's also quite interesting that we live in a very complex environment, and there are many different types of entropic processes going on at all times. And depending on which parts of our environment we focus on, we see more entropic processes going on, more positive entropy, more things running down, or other parts of our environment we may not pay any attention to. So we don't see the entropy in those parts of the environment. Time may run differently for different observers, as Van pointed out.

Now when I moved my own career from electrical engineering to biology, my favorite book was a book entitled *What Is Life?*, written by somebody who was anything but a biologist. He was Erwin Schrödinger, one of the greatest physicists of the last century. In this little book, *What Is Life?*, which you can always find on Amazon, and it's very small, not even sixty or seventy pages, I think, Schrödinger comes up with a different definition of life than any biologist had ever come up with. Schrödinger said what life is is a process of *negative entropy*; instead of things becoming more disordered, things are becoming more ordered pursuant to a code. A reproducible or, in his words, a "periodic" code. For example, DNA is such a code, and instead of us becoming more disordered from the time we form a zygote, we in fact become more and more ordered. And so he looked at life as something that went, for a period of time, quite contrary to the direction of the second law of thermodynamics. Now this idea of a code, in my mind, was absolutely beautiful, and I locked on to it because it seemed to be the pathway to solving a lot of diseases, as D.A. mentioned at the beginning, which is what my day job is all about.

In the course of solving these diseases, I came across another wonderful book written by another individual who had not a bit of training in the life sciences, Ray Kurzweil. Ray Kurzweil wrote a book, *The Singularity Is Near: The Age of Spiritual Machines*, and in all of his books he puts forward the notion that we are gradually compounding in our code computation; our codes are gradually becoming more and more sophisticated, able to compute and decode greater and greater amounts of entropy; and, with that, the current rate of exponential growth in computation and information processing in just a few decades' time, certainly in just a few centuries of time, we will have a computational capability equal to that of all of the atoms in the entire known Universe.

-123-

So Ray Kurzweil posits that the Universe will not end in a cold end of time, total state of entropy, as I mentioned at the beginning, nor will the Universe end in a big, massive, gravitational, heat-oriented crush, but instead, our growing intelligence from this compounding of knowledge and information processing, will allow us to discern new laws of physics that are not currently understood today. Things, laws of physics, and layers of nature that more deeply underlie physics today, just as our physics of today more deeply underlies reality than did Newtonian physics. So in Ray's view we will very likely end up streaming order back into this Universe from some other universes in the multiverse that Van referred to previously to enable an indefinite, infinite extension of biology in this Universe.

Now, that, finally, links with my first mentor as I was coming out of college, and in fact the main reason why I feel so excited to be here at the first annual InterPlanetary Festival today. And that individual is Gerard K. O'Neill, who propounded the idea in the 1970s and 1980s that the natural place for an expanding industrial civilization, this type of Kurzweilian civilization, would be building Dyson spheres and things like this, is not on the surface of a planet but is instead out in the cosmos. That we should have a cosmic consciousness living out in the cosmos, and within just a few decades' time, there would be more humans living off the Earth in space habitats at Lagrange stable gravitational points between the Earth and Moon, known as L4 and L5.

I found this idea absolutely inspiring and completely logical, and have been devoting my life to it ever since. And I'm so glad to see that the Santa Fe Institute is bringing all of its great intelligence and insight into reality to help serve as the key enzyme to push human civilization out off the planet and into the cosmos. Now, as Gerry O'Neill reached the end of his life, he began to grapple with just how long it would take to move us all to be interplanetary, and he wrote a book called *2081: An Optimistic View for the Human Future*, which envisions the world in 2081, in which there would be more people living off-planet than on-planet, many of them in these vast habitats of Bernal Spheres around the L4 and L5 Lagrange points.

O'Neill grappled with a simple question. He sat there and he spent something like five years reading every science fiction book he could read, every scientific forecast he could read, from the expert forecasters throughout the twentieth century. And he said, "Why are people almost always wrong about what next technology is going to be available in ten years or twenty years?" And "Why are people almost always wrong with how soon the technologies will come?" As an example, he said, "Where is the flying car?" Since the Jetsons in the '60s we've been waiting for the flying car. We were wrong. It wasn't around the corner. On the other hand, if you take a look at the most far-sighted science fiction writers in the twentieth century—the Robert Heinleins, the Arthur C. Clarkes—none of them saw that billions of human beings in just a few decades would have in the palm of their hand the knowledge of the entire world, essentially for free. So he figured that everyone having knowledge in the palms of their hands was something that would occur distantly in the third millennium, 2200s, 2300s, *Star Trek* days. But no! It came around in 2005.

So why this difference in perception? I think the solution to this lies in the idea that Van and Stefani identified previously: that we humans have a big effect on the actual passage of time; that it is a function of perception; that when we build more order into the world, as Schrödinger said, we actually call the future forward to us; we pull the future forward with our own activities. So all of you being here at this InterPlanetary Festival, you are helping to create time, you're not using time, and you're bringing the best of times forward to today.

APPLAUSE

**S. CRABTREE** I just want to say one last thing with that in mind, Martine. Humans are nothing if not inventive, and one of the things we can look at in the past 200,000 years of *Homo sapiens sapiens* is how we can imagine our futures. I think that that is exactly what's happening. If we look archaeologically, we can see people imagining their futures and bringing future to them, and that's what we're doing here today. So we need to be more imaginative, and we need to think about time and our ability to change it.

-125-

Do we have time for a question before we're done? Should we take a couple of questions from the audience before we go?

**AUDIENCE MEMBER** I want to ask a question, okay? But I can't talk to you that way.

.....................................................................

## We actually call the future forward to us; we pull the future forward with our own activities.

.....................................................................

**V. SAVAGE** Is there a microphone to take around? If you stand over here, I can repeat it to the audience.

**AUDIENCE MEMBER** No, no, no. I want a microphone. You have in your hand a microphone. I want one. Okay, can you hear me?

**V. SAVAGE** We can hear you. Go for it. You're good.

**AUDIENCE MEMBER** Hello? Okay. Well, let me introduce myself. My name is Boris Bluebeard, and I am a nuclear engineer, I used to be an engineer working in CERN, and I'm an artist, diplomat, warrior, and a lot of other things. I speak fifteen languages, and I've been to twenty-eight countries, but now I'm very excited to be familiar with Neil deGrasse Tyson.

My first question is why he's not here to give us some laughter about the logic of human beings. The difference between chimpanzee and human beings, it's just one percent of DNA. He said, "Do we know too much?" but I don't think we know too much at all. This is what he said: Just imagine that another species, they can be anybody from any kind of multiverse, they can consider us out of line with everything. So they can say okay, you know, but astrophysicists . . . well, you heard about Neil deGrasse Tyson?

**V. SAVAGE** Yeah.

**B. BLUEBEARD** The question is this: How far we can go in the future to expect to know—you know what, we can expect too much—but how far we can go to go to Mars? How far until we go to Mars. I'm ready to go!

**V. SAVAGE** To go, or to live there?

**B. BLUEBEARD** To live there and stay.

**V. SAVAGE** Okay, I'll start answering that. To go is very feasible, actually. Not easy, but I think it's definitely within our reach. It's a much harder question of how we're going to live there and stay there. One thing we didn't talk about, but Stefani mentioned beforehand, is that in some sense our planet is changing. Climate change is changing our planet into a different planet. You already see flaws of timescales, with insects or plants coming out at the wrong time, things not eating them, species can't coexist. So we need to solve those problems in some sense in the same way we need to think about how we could have life on Mars.

And while those challenges, figuring out those syncing cycles, are things that our ancestors have done in the past, we're going to have to figure out as we do colonize Mars, if the plants that we grow, are they going to grow? Is corn going to be 120 frost-free days like we have here, or is it going to be something different because it adapts to the ecosystem that we build there?

So these are challenges that we have to adapt toward, why we need models that are built on the past, and why we need to be flexible into the future.

Thank you. Okay. We'll be on the side if anyone wants to ask questions, and also there's a book signing when Martine comes out, which everyone should go to.

APPLAUSE

# INTRODUCTION:
# MOTION & ENERGY TECHNOLOGY

Perhaps the most self-evident challenge to the interplanetary vision is that of space travel. How will we get from here to there? The Apollo missions of the late 1960s and 1970s are the farthest humans have been from the Earth's surface so far. In Solar System terms, the Moon is just around the corner—we haven't even left the block we grew up on.

A few basic calculations show extra-solar travel—that is, travel beyond our own Solar System—to be a daunting task at best. The closest to our own is the Alpha Centauri star system, which is over four light-years away. A voyage at the daunting speed of a tenth of the speed of light would take forty years one-way, and a round-trip would literally be the voyage of a lifetime. Basic energy calculations show that even this speed is well beyond our ability, and so it appears that any mission will need to span multiple generations. Travel within our Solar System—to Mars, the asteroid belt, or farther—seems eminently more feasible, and yet these destinations appear at first glance to be impossibly challenging, too. We need technology to land large payloads on the Martian surface. We need to prevent radiation from giving voyagers cancer or causing irreparable damage. We need to know how to stop the spread of disease in a vessel with no fresh air and the nearest hospital literally millions of miles away. We need to understand human interactions well enough to maintain cohesion when there is nowhere to get away and vent.

And yet, there is reason for optimism. We are in the middle of the next great revolution in space technology. Miniaturization of sensors and improvements in computational power mean we can now do more with a one-kilogram payload than we could previously do with a thousand-kilogram payload, and, if trends continue, we'll be able to do a lot with a satellite weighing only one gram. Simultaneously, there has been a significant reduction in the cost to get to space, driven by a new wave of private space companies supported by governmental space institutions. It may not be long before an orbital trip around the Earth is a luxury that can be purchased. We understand better than ever how to stay alive in space and the impacts of a microgravity environment on the human body—there has been at least one human in space continuously since the year 2000.

These discoveries inspire us to keep up the human drive to explore, and provide increased hope that we may finally prove we are not alone in the Universe.

Concurrently, there has been a meteoric rise in our understanding of exoplanets, that is, planets orbiting a sun that is not our own. In the beginning of this millennium, there were about thirty known exoplanets. By the time of this panel, over *three thousand* were known. These discoveries have radically changed our understanding of our galaxy. It is now believed that it is common for solar systems to have one or more planets, and Earth-like planets are being discovered every year. The TRAPPIST-1 star system alone has at least three planets within the habitable zone, where liquid water could exist on the surface of the planet. These discoveries inspire us to keep up the human drive to explore, and provide increased hope that we may finally prove we are not alone in the Universe.

The task of this panel was to discuss this past looking toward the future, and to bridge the gap between dreams of exploration and the reality of the daunting challenges ahead. Fortunately, I had two amazing panelists with me to address these issues. Plus, what are festival panels for, if not to be overly ambitious?

The first panelist is Pete Worden. Pete has one of the most interesting and diverse careers in all of aerospace. He received his PhD in astronomy from the University of Arizona and additionally worked as a research professor there. He has an impressive scientific publishing record, but that is only the beginning of his accomplishments. He served in the Air Force, retiring as a brigadier general after twenty-nine years of active service. Among his many accomplishments, he was integral to the Strategic Defense Initiative, having twice served in the Executive Office of the President, commanded the 50th Space Wing, and served in the Air Force Space Command. I met Pete while I was at Stanford doing my PhD, and he was serving at the highest level of the civilian space program, directing NASA Ames Research Center. Pete now has what he described to me as "the best job in the world," working as the chairman of the Breakthrough Initiatives. He is responsible for the creation of the Starshot mission to send a probe to Alpha Centauri. This project is the embodiment of "you're working on the right topic when others say it is impossible."

Our second panelist is Neal Stephenson. I first learned about Neal when I was in high school and a reader of science fiction. I remember browsing at the local bookstore and being drawn in by the intriguing title: *Cryptonomicon*. He quickly joined Isaac Asimov in my mind as someone who had thought deeply about using storytelling to explore the impacts of technology. I was far from alone in holding him in high esteem. He is well known for building worlds that address big ideas, be it the interplay between the physical world and the virtual in *Snow Crash*, futuristic monks interfacing with the many-worlds interpretation of physics in *Anathem*, or exploring a future where the Moon has split apart in

*Seveneves.* It is important to note that he is not just a world builder but is also deeply engaged with understanding the science behind his books. I first met Neal over tea at the Santa Fe Institute, where we had a conversation on the particulars of orbital dynamics and the consequences for exploring the outer reaches of the Solar System. Neal has also worked in the space industry, most notably during the early years of Blue Origin's existence.

It is with these two that I had the privilege to discuss the future of humanity as a space-faring species. They share a love of space and a desire to see big accomplishments and radical ideas. They are also split, embodying the differences of fact and fiction, realist and dreamer, optimist and pessimist. It may surprise you to learn who is who.

—*Brendan Tracey*
*DeepMind*

## MOTION & ENERGY TECHNOLOGY

*D.A. Wallach introduces the panel, moderated by Brendan Tracey
and featuring Neal Stephenson and Pete Worden.*

**D.A. WALLACH** All right, everybody, moving right along. We are
now going to talk about two more critical aspects of the interplanetary life support systems that we need to consider, and these are
those having to do with energy and motion. How are we going to
create the energy needed to explore the broader Universe once we're
out there? How are we going to get back? How are we going to
sustain ourselves in these foreign territories? And how are we going
to get around within them?

We struggle to move ourselves around this planet, in cities, between
continents, across oceans, but we're going to be dealing with landscapes that are completely different, which will call for new technologies. We couldn't have a better group equipped to discuss this.
First, Neal Stephenson joins us. Neal is one of the most acclaimed
authors of speculative fiction alive today. Neal's work defies cate-
gories; it has been branded science fiction, historical fiction, cyber-
punk, post-cyberpunk, and even baroque!

LAUGHTER

**D.A. WALLACH** But it's all interesting, and he's going to lend
his singular perspective to thinking about these questions that may
turn from speculative to practical in short order.

Brendan Tracey joins us from the Santa Fe Institute, and Brendan
uses statistics, machine learning, and engineering analysis to

evaluate the designs of physical systems. So he asks—for systems that are too expensive for us to play with in reality—whether we can shortcut our understanding of them in the digital domain. Another skill set that presumably is going be useful for imagining a context that we can't experimentally probe that easily or cost effectively.

And, finally, Pete Warden joins us from the Breakthrough Prize foundation where he is executive director and runs the foundation's Breakthrough Initiative. Pete has an incredibly decorated past, from the United States Air Force to academia at the University of Arizona, where he's a widely published astronomer, and more recently as director of NASA's Ames Research Center at Moffett Field, California. So please give it up for our panel, and we'll see you in a few.

APPLAUSE

**BRENDAN TRACEY** Great. So, given that we're here at the InterPlanetary Festival, we have a couple of good planets in our Solar System here, but what do you two think: Are we ever going to see humans setting foot on another planet outside of our Solar System? Or do you think we'll be limited to probes? Or do you think we're not going to get there at all?

**PETE WORDEN** Well, maybe I can start. A lot of it depends on what you mean by *humans*. I think absolutely. In fact, I always thought the Solar System's pretty boring.

LAUGHTER

**P. WORDEN** There's like no scary aliens or anything, but we now know that essentially every star in the galaxy has a solar system. That at least a quarter of the stars, like the Sun, have a planet at least the size of the Earth in the habitable zone. So I think that there are a lot of places to go, and inevitably we will expand there. The question is, what do you mean by *we*? I think we're on the verge of becoming a different species either through our biological manipulation or integration with AI. And I suspect that the easiest way to settle a planet—say there's a cool planet in the Alpha Centauri system, which there probably is—would be to send a small probe there that then could receive, at the speed of light, signals and print out or boot up a civilization. So I think that's going to happen at the end of this century.

**NEAL STEPHENSON** Yeah, I guess I agree that it all comes down to how we define *we*. If *we* is people like us, my personal opinion is that we're never going to leave the Solar System, with one possible exception that I'll get to in a minute. But the energy requirements, the transit time, the risks involved are so enormously greater for interstellar travel as compared to interplanetary travel that a lot of things have to change in order for living organisms to make that trip. The one exception that I could see would be—it's something that's raised in the science fiction show *The Expanse*, where there's a religious group that decides they want to go on an interstellar mission. And I think that is the only circumstance under which humans like us would ever actually make that trip because otherwise it is so expensive and so impractical. It would take such a long time that I think that the only thing that could motivate it and cause humans to put out that much effort would be some kind of intense religious or ideological conviction.

-135-

......................................................

# I think we're on the verge of becoming a different species either through our biological manipulation or through integration with AI.

......................................................

**B. TRACEY** So both of you mentioned maybe tweaking what we mean by *human* in order to get to other places, and there are a couple of ideas, right? So one is that we could send maybe a pill capsule or maybe something larger that has an ecology of bacteria in it that could grow. Or there's another idea of von Neumann machines that can replicate themselves and possibly boot up to the next solar system to land on new planets. Do you think those are more fruitful avenues for us to be exploring today in order to meet these goals? Or do you think different kinds of technologies are more appropriate?

**P. WORDEN** Well, to go back to some of the discussions from the previous panel, predicting the future is risky. I think we're on the verge of having many technologies that will change what we mean by *entities*. But it strikes me that, again, a planet orbiting a nearby star may have all the raw materials, and may be an ideal location to take the resources there and use them to boot up something. Now, is that something biological? Or is it a computer? Or a von Neumann machine? It's hard to say, but clearly it would have to be able to replicate itself or to make something that could replicate itself. We call that *life* today, but it could be something more complicated.

There are almost no people in the space business I know who weren't inspired by the literature they read in science fiction.

**N. STEPHENSON** The most compelling description of this that I've heard is from Freeman Dyson, who has suggested that it might be possible to biologically engineer plant- or tree-like things that would grow in the Kuiper Belt or the Oort Cloud, that would basically live on snowballs extremely far away from the Sun and well beyond the orbit of Pluto. And if you could create a seed, and plant it in a snowball, it would actually grow into something like that and make more seeds. Then over time you could see that propagate outward from the Sun and eventually cross the boundary into the Oort Cloud surrounding the next star over. In that way this sort of forest of living organisms could very gradually spread across interstellar distances and support all other kinds of life-forms along with it, including life-forms that might be descendants of people sitting here in this audience.

**B. TRACEY** So, Pete, you mentioned that it's really hard to predict the future. One quote that I've really loved is "History doesn't repeat itself, but it does rhyme." I was curious to know what you two think about how much our past histories of exploration—the motivations of those and the stories that we tell about them—how much those are informative as to the future of, say, intrasolar travel, within our Solar System? To what degree are those stories relevant, and to what degree are those stories worthless because technology is so different and the times have changed?

**P. WORDEN** Well, I don't know. I mean, there are almost no people in the space business I know who weren't inspired by the literature that they read in science fiction. I remember in high school I used to hollow out my history book to put a science fiction paperback in, but it didn't fool anybody. My mother told the teacher, "Leave him alone. He won't bother anybody if you let him read." And I'm sure there are others that are similar.

**N. STEPHENSON** Yeah!

**P. WORDEN** It's critical, I think, that we have this past that prepares us for the future. And so I think that it's a continuity. It's certainly not something new. No matter how right or wrong the science fiction is, it puts us in the right mindset, I think, for expansion into the Solar System and eventual interstellar expansions of some form.

**N. STEPHENSON** I think we're going to see what amounts to different cultures arising around . . . and this is already sort of happening in the commercial space world, where you've got some people who are serious Mars advocates and others are saying, "No, let's go to the Moon." There are others who want to use *in situ* resources to build habitats, etc. So I think that different discernible

-137-

cultures are going to arise around those different approaches to moving into space and that the kind of person who thinks it's a good idea to gather a bunch of rocks together to make a rotating habitat is just a different person, culturally and psychologically, from someone who wants to go to Mars, or someone who wants to go to the Moon, to say nothing of the idea of making the much more difficult leap into a neighboring solar system. I think that there's a lot of fodder there, and, again, this has been touched on already in *The Expanse* where there are distinct cultures associated with the Earth, Mars, the belt, and so on.

**B. TRACEY** So, Neal, you mentioned private companies, and one of the biggest differences now in today's space world versus how it has been is this shift between governments making a lot of the progress to more and more private companies making these big groundbreaking things. Both of you are very interesting in that you've both been involved with the private space sector as well as the governmental side of things. I'm curious to know whether you think that this shift to a private space is inevitable and where we're headed, or if you think the shift is going to come back? And, secondly, do you think that's a good thing, whichever way is the way things seem to be heading?

**N. STEPHENSON** Well, I don't think we should ever discount the fact that the private space companies couldn't lift a finger without more than half a century of government-funded research that told them how to do it. However, having said that, I think that the advances that particularly SpaceX has made in the last few years are almost unbelievable in the speed with which they've been able to do things like land reusable stages on barges and hit other milestones that I suspect would have taken a lot longer under a government program. Pete might even have opinions about that.

**P. WORDEN** Well, I spent most of my career working for governments. I love governments, if anybody's listening—

LAUGHTER

**P. WORDEN** —but I tend to think that it's really a public-private partnership.

**N. STEPHENSON** Yep.

**P. WORDEN** And we're moving more to the private side of that. I mean, I've worked for NASA for ten years, and its predecessor agency was called the National Advisory Committee for Aeronautics, which helped build the industry. They didn't actually build many airplanes. Now, I have to give a short advertisement to, well, in addition to the Breakthrough Prize foundation, I'm a consultant for the Grand Duchy of Luxembourg, which is an unlikely space, a public-private partnership, but they have started a major effort for space resources, and it's sort of the quintessential public-private partnership because they have sovereignty and are willing to use it to help industry. And they're also working to get global buy-in for the idea of using space resources. In fact, the deputy prime minister would say that the language, at least, spoken on the Moon and Mars is Luxembourgish. I don't know any, but—

**N. STEPHENSON** You're referring to the partnership with Planetary Resources?

**P. WORDEN** And others.

**N. STEPHENSON** And others?

**P. WORDEN** They've signed about twenty different agreements, and there's more coming.

**N. STEPHENSON** They've been busy, huh?

**B. TRACEY** So, if I could speculate for a second on this, one of the things that's really interesting to me is that a lot of the private space companies seem to be part company, and part vision. I did my PhD at Stanford in aeronautics and astronautics, and I have a lot of friends who have gone into that industry. One of the biggest things that attracts them to these companies is the existence of true believers in the company. That is, the people who realize that they need to make money to keep doing what they're doing, but that's not really why they're doing it. Why they're doing it is to continue this space mission. For another example: Elon Musk has said that he wants to retire on Mars. I'm curious to know if you think those kinds of visions are really what is driving things forward, or if at some point it's going to have to become a money business and the culture will have to change?

**P. WORDEN** Well, it's clearly both. Successful entrepreneurs usually have a vision of a future, a utopian vision, and they try to make it real, but they need resources in order to make it real. I think Jeff Bezos is the key example. He told me when I first met him that he was selling books so he could build rockets.

LAUGHTER

**P. WORDEN** And so I buy his books. But I think that there's a spectrum. The foundation I work for is purely philanthropic. We're trying to do interstellar stuff, to look for life. On the other extreme, there are those who are pretty heavily motivated by the money. I think that all of it requires this inspired vision. They all share the vision that humanity should be expanding into this Solar System and beyond.

**N. STEPHENSON** Yeah. I think if your sole objective is to maximize profit, then you're not going to build a space business. But there are plenty of people who wanted to go into space who don't have the wherewithal and the business acumen to make it into a sustainable thing. So what's going on right now is that there's a generation of people who do have the knowledge and the wherewithal who were raised to believe that by the year 2000 we'd have thousands of people living on Mars. When that didn't happen, I think a lot of them kind of realized that it was up to them now.

**B. TRACEY** Do either of you have opinions on what you think the "smoking" of space is? That is, something that we're all doing now today but we might realize in twenty or thirty years that maybe it wasn't such a great idea. That maybe we should have been doing things differently?

**N. STEPHENSON** Smoking?

**B. TRACEY** Yeah. It was widely accepted for a while. It was a giant cultural thing, but now we've learned that there are negative repercussions. And I'm wondering, what's something like smoking that we're doing now in space, that we're going to look back on and say, "Oh, we shouldn't have been doing it this way. We should have been doing it this other way all along"?

**P. WORDEN** I'll take a stab at it: chemical rockets.

**N. STEPHENSON** You stole my line.

LAUGHTER

**B. TRACEY** Aha, so we have agreement! Well, what would you replace it with? Maybe you two disagree there.

**P. WORDEN** Well, that's hard to say. I mean, I'm particularly a fan of directed energy, of keeping the fuel behind and beaming it to the vehicle to get it off the planet. I think there are other methods as well, but chemical rockets are, well, they just barely work. So I think there is some alternative. Obviously, if you can harness nuclear fission or fusion in some way—and there are a few challenges there, to be sure—those are the two areas as I see it. Beaming the power either through lasers or microwaves, or a much more efficient conversion of energy to specific impulse.

**N. STEPHENSON** Yeah. There are a couple of promising approaches to the directed energy thing. Lasers and microwaves have both been looked at in a lot of detail. There are detailed plans on how those kinds of systems might be built. The funny thing is that once you start looking into this, you find that really smart people have been for, like, a hundred years dreaming up alternative ways of hurling things into space, but there's basically nothing that you can come up with that wasn't already invented by some dude in Russia sixty-five years ago.

LAUGHTER

**N. STEPHENSON** So you just have to look at what's out there. The late Jordin Kare was a passionate advocate for a laser-based directed energy launch system, and one of the comments he made was that amateurs think about specific impulse, which is to say the performance of the rocket, and pros think about insurance.

LAUGHTER

**N. STEPHENSON** You can invent all of the cool new propulsion devices you want to, but, unfortunately, sooner or later you've got to make it an insurable thing. So a lot of the weird stuff we see, like Elon Musk launching a car into space, seems like a completely bizarre thing to do until you realize that he would have had to do it anyway with a bunch of cinder blocks in order to prove that the rocket works, just so he could buy insurance.

**B. TRACEY** So there's a company that you might not know called Planetary Resources, and their goal is to mine asteroids. And I was at a talk by them when they were just getting started, and one of the things that they pointed out is that mining asteroids isn't necessarily about bringing gold from an asteroid back to the Earth. What it's about is being able to mine fuel on asteroids. And this means that, rather than having to take fuel and launch it from Earth into space, you can grab fuel from asteroids in space, and that makes access between different Solar System bodies much cheaper. To me, this was a great idea. Now I'm learning chemical rockets may be bad—maybe we'll see—but now I'm wondering if there are other game-changing technologies that you're really excited about in terms of access to space and to our Solar System.

**P. WORDEN** Space, at least in the inner Solar System, is flush with energy. And I think that whether it's volatiles that you mine on the Moon or asteroids, using resources you find there, that converting energy to propulsion is probably the next step. And, once you're in space, you're out of the Earth's gravity well. Then it's just a matter of energy and material. As far as an electric propulsion system or other kinds of things, I suspect that what we'll do to use the volatiles we'll find, such as water, is not to burn them in space but we'll use the ample electricity to get much more efficient propulsion. An electric system could be very efficient.

**N. STEPHENSON** Yeah.

**P. WORDEN** Although the metal stuff is a pretty neat thing if we can make it work. We're kind of using up a lot of platinum group metals that maybe we can change in the future, different technologies, but at some point it's going to be cheaper to bring that stuff from space than it is to dig down three kilometers into the ground to get it.

-143-

**N. STEPHENSON** Yeah. It's about reaction mass, as I'm sure you understand. A rocket is weird because it's doing two unrelated things at once. It's burning two chemicals to make energy, and then it's throwing stuff out the back as reaction mass. And it just so happens that those two are the same stuff, but they don't have to be the same system in that way. And once you get out of the gravity well where, as Pete says, energy is a lot more plentiful, you can decouple those activities and come up with new ways of getting around.

**B. TRACEY** Another really exciting thing about the interplanetary mission is finding extraterrestrial life. And one major effort in this is the Search for Extraterrestrial Intelligence, SETI; they mostly look at radio waves and try to find interesting signals. Do you both have any crazy ideas for what other kinds of signatures we should be looking for, for finding life? If you had all of the resources you'd ever want, you could build some interesting thing that detects a weird pattern, maybe.

**P. WORDEN** Well, I think as technology advances . . . there was a seminal paper in 1959 by Morrison and others that suggested that radio-frequency energy at around one to ten gigahertz would travel through the interstellar medium at great distances across the galaxy. We were using radio, so it would suggest we should look for radio. Today we're using optical frequency and laser communications. In fact, I understand that the country of Norway is going radio quiet in the next few years just because they're going to all light transmission and cables and so forth.

**N. STEPHENSON** Oh?

**P. WORDEN** So I would say whenever we have a new idea. In fact, one of the things my foundation does is we're looking at optical and every other frequency we can look at. People have suggested gravitational waves, although right now the only way to make them is to take a couple of solar masses and convert them to pure energy and disturb the space–time.

**B. TRACEY** You could build a few black holes.

**P. WORDEN** That's a little too expensive to communicate.

LAUGHTER

**P. WORDEN** I don't know. There's some pretty speculative stuff, which physics says doesn't work, but there are people who are talking about maybe doing large-scale quantum entanglement and preserving that over long distances. It violates several laws of physics to say that it communicates, but, again, nobody knows what an advanced technology would do, and it may be right under our nose. I think we do the best we can with what we understand.

**N. STEPHENSON** I just remember a quote from Bob Zubrin at Starship Century [Symposium] a few years ago. Somebody asked him basically this question: "Could we detect alien spaceships by looking for some kind of radiation coming out of their drive systems?" And he just said, "That would be like trying to find a Formula One racer by looking for the heat signature of his whale oil lanterns."

LAUGHTER

**N. STEPHENSON** It's funny, but the point he's getting at, of course, is that, in order for us to answer that question, we have to anticipate technologies that we haven't thought of yet and then try to imagine what the signature coming out of those would be. And that's why, again, to go back to Freeman Dyson, that's how we get

Dyson's spheres. It's like, "Okay. We can't predict specific technologies, but we know certain facts about thermodynamics, and so what are some things we could look for that would have to be true regardless of what technology was being used?"

**B. TRACEY** One thing we were kicking around at lunch at the Santa Fe Institute one day was that one technology that's coming is to be able to look at the atmosphere of extrasolar planets, which is, to me, totally bonkers that we can have something so far away, and so dark in the sky, and yet we can measure the signatures. And so one thing that we might try to measure is rapid fluctuations in these signatures, possibly. And especially if we look at ourselves as humans and other biological examples where when something new starts, all of a sudden there are these massive measures of inefficiency. And then evolution kicks in, and other pressures kick in, and things get more and more efficient, and the signal dies. And so that's maybe another thing that we could think about—how to look for those signatures of inefficiency, because maybe that's a sign of things changing a lot.

-145-

································································

There are lots of TV shows and movies where the aliens come to our planet to get our water or something. That's a long way to go to get some water; it's a lot of trouble to go to. And I don't think we have anything that's worth covering all that distance to get.

································································

And so, to flip the question around, many have said that we should not be trying to broadcast our existence because really bad things might happen to us as a consequence, but let's say we're totally going to ignore those people, and we want to do the best we can to alert the galaxy that we're here. How would you go about designing that project?

**N. STEPHENSON** I guess I'm one of those who's not worried about it. I don't see any reason to hide. There are lots of TV shows and movies where the aliens come to our planet to get our water or something. That's a long way to go to get some water; it's a lot of trouble to go to. And I don't think we have anything that's worth covering all that distance to get. I'm not worried about it. As to how to broadcast, again, it's a reversal of the earlier question. We can think of things like broadcasting radio waves, which we're doing anyway, but would another civilization think of that? It's hard to know.

**P. WORDEN** Probably the best way we know today is a very powerful laser. And I think within a decade we're going to find a number of potentially life-bearing planets around nearby stars, so it's reasonable to maybe aim a laser at them. I mean, obviously in the far distant future you could—okay, take another science fiction thing: *Star Wars* had something called the Starkiller. Maybe you could fire a set of stars off. Now, that's pretty wasteful, and you would hope there's no life in those systems. But I think as we develop new technologies, if you want to try to communicate, that's the best way. Now, I have to be a little nuanced on this. In the foundation I run, our principal science advisor was Stephen Hawking, and he had this rather negative view on communicating, so I'm not supposed to have an opinion, but I tend to agree with Neal.

**N. STEPHENSON** And just the back half of that is, if they're capable of coming here, we don't have much of a chance anyway. So we might as well just—

**P. WORDEN** They're probably already here.

**N. STEPHENSON** Yeah. Yeah.

**B. TRACEY** So it's time to wrap up. I'm wondering if either of you have anything that you'd like to say to the festival before we wrap up.

**P. WORDEN** Cool festival!

**B. TRACEY** Ha, excellent. And just to let everyone know, Pete is going to give a talk in a couple minutes about the Starshot initiative and some other things he's working on, which should be really great. And Neal will be doing a book signing, as well. You should all attend both events and thank you for listening.

**N. STEPHENSON** Thanks. Yeah. Thanks in the front.

APPLAUSE

**P. WORDEN** Thank you.

# INTRODUCTION:
# LIVING IN SPACE

Humanity has long glamorized the possibilities of living in space, of establishing a small colony somewhere beyond Earth, where the advances of science and engineering allow us to explore the limits of a new frontier.

While these dreams have always been filled with aspiration and pushing the realm of the possible, actually living in space comes with a host of realities and challenges that are not easy to solve. Nor is it easy to find human subjects willing to be subjected to the realities of such an isolated journey.

Solving these challenges is, in a way, at the core of why space is so invigorating: it begs the creative mind. While an interplanetary existence is intriguing, we are also reminded that, even in the technologically advanced world of 2018, we do not fully understand how we might comfortably, or practically, live in unforgiving areas of planet Earth such as Antarctica or the Sahara Desert.

As we seek out the rare souls who are willing to pioneer such solutions, be they here on our own planet or somewhere beyond in the cosmos, the greatest value we find in such research and exploration is a better understanding of ourselves. Going into outer space requires a fundamental understanding of what works here on Earth, whether it be in human connection psychology or embracing unique engineering solutions for enduring habitats.

SFI gathered a panel of men and women who have devoted their lives to solving hard problems and being prepared in (and for) the extreme, with the importance of teamwork linking their seemingly disparate roles.

Beginning with Ashton Eaton, a two-time Olympic gold medalist and world record-holder in the decathlon, this discussion provided a deep perspective on how one trains for a distant goal requiring extreme focus in order to achieve beyond one's own limits. Ashton touched on "human factors"—the senses and emotional connection to teammates—that led him to extraordinary performance in the Beijing and London Olympiads.

In Nina Lanza, this discussion had an accomplished scientist who not only designed some of the sophisticated technology for the Mars Curiosity rover but actually lived in Antarctica for eight weeks as part of a NASA experiment to understand team interaction in a simulated Martian environment. She reflects on the realities of living in an eight-by-eight-foot tent with few reprieves into the barren cold of Antarctica.

-149-

Haym Benaroya has dedicated his life to the very challenging yet practical question of living in space: Where do we live? Or, more specifically, what structures might we live in? The architecture needed to protect against radiation and the unforgiving environment of the Moon or Mars means engineering small, austere structures that have little resemblance to the homes and habitats most societies have come to enjoy in Earth's open, oxygen-saturated climate.

Finally, I've spent a career working in teams across national security and special operations. My work has been in understanding the first principles of high performance—or, put simply, the elements of teams that are transcendent across the military, medicine, and sports.

This discussion covered a wide range of topics. If nothing else, it was an important reminder of two realities: First, although we have amazing scientific and technological capabilities, the challenge of actually colonizing beyond Earth remains relatively distant but is buoyed by the continued advancements of people who SFI had gathered. And second, even in beginning to solve those challenges, as Haym Benaroya and Nina Lanza are, the human being is still at the core of success. Finding men and women who are able to master their own minds and do so with teammates traveling together in a small space for years—well, that's a bold human endeavor in and of itself.

Although the panel was candid in discussing the daunting realities of living in space, it was also an inspiring reminder of how unique our lives here on Earth are, and how much the frontier of space allows us to understand about the unique existence of human beings on this planet.

—*Brian Ferguson*
*CEO, Arena Labs*

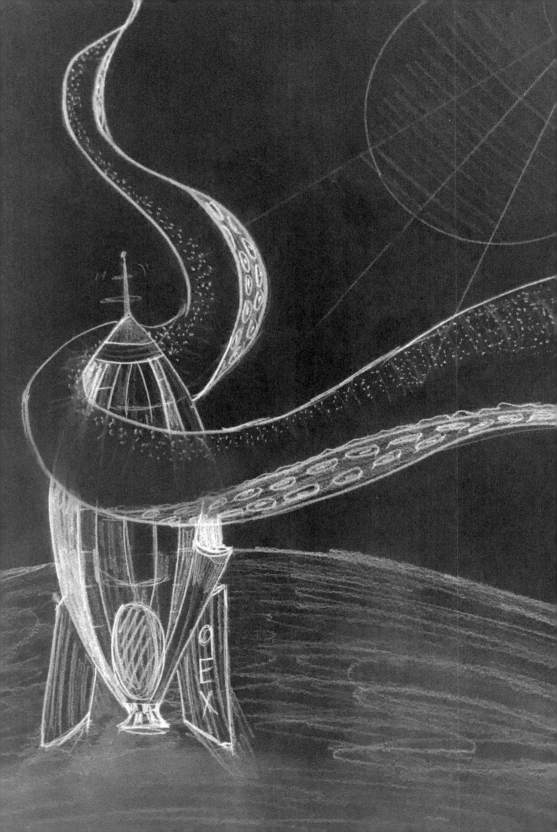

## PANEL:
## LIVING IN SPACE

*Sandra Brice and D.A. Wallach introduce the panel,*
*moderated by Brian Ferguson and featuring Haym Benaroya,*
*Ashton Eaton, and Nina Lanza.*

**SANDRA BRICE** Good morning, everyone, and welcome to day
two of the InterPlanetary Festival. I'm Sandy Brice, director of
events for the Railyard, and I just want to say thank you again
to SFI, to Caitlin McShea, David Krakauer, Kayla Savard, Tim
Taylor, and the entire SFI team. You guys have just been tremendous in putting on a very ambitious project for the first time, and
we hope it's a long and glorious relationship with the Railyard.
We love having you here.

I also want to make sure those of you who are here for the
InterPlanetary event this morning know what else is going on. This
event is a wonderful collaboration with CURRENTS New Media.
They're putting the final touches on their installation as we speak
inside of El Museo Cultural right down here across the tracks. If
you've never been to CURRENTS, I promise you, you're in for a
mind-blowing treat, so please go tonight. They open at six p.m. and
they're open until midnight. We'll have a fabulous concert here
tonight with Max Cooper on the InterPlanetary main stage, co-
presented by SFI and Meow Wolf. And the conjunction of these
events happening at the same time, in little Santa Fe, New Mexico,
is nothing short of extraordinary. So please go see world-class media
art of all kinds from all over the world tonight, inside El Museo.
Anything else, Caitlin?

**CAITLIN MCSHEA** Yes, CURRENTS is the official InterPlanetary afterparty, so you have to go!

**S. BRICE** CURRENTS is the official afterparty for InterPlanetary. So there you go, you have to go. Okay. I'm going to turn over the day's events to D.A., who's going to tell you everything that's about to happen.

**D.A. WALLACH** Thank you very much. Welcome everyone! We're going to get right into it because we're running a few minutes behind, and I don't want to take time away from this extraordinary panel. I'll just give you a couple words about each of the panelists. Haym Benaroya is a professor at Rutgers University in the Department of Mechanical and Aerospace Engineering, and he focuses on structures that are capable of enduring extreme environments—of which space is obviously a great example.

Nina Lanza is a staff scientist at Los Alamos National Labs. She told me the other night about her favorite gemstones, which are labradorite and tanzanite. Of course, tanzanite can only be found in Tanzania. And she works on the history of water on Mars—she focuses on manganese in the Martian environment. So get ready!

Brian Ferguson, whom I've known for a couple of years, was a Navy SEAL and now coaches all sorts of organizations on high performance.

And finally we have an Olympian, Ashton Eaton, here who holds the world record in the decathlon. I need say no more. With that, give it up for the panelists. Thanks!

APPLAUSE

**BRIAN FERGUSON** Good morning. All right. Thank you guys for your patience. This, I think, is almost a tease. You can see the panelists I'm super humbled to be up here with, and we're going to talk about the idea of living in space. What does that entail? What goes into planning? We could easily spend thirty minutes just on the bios of each of the folks up here, so we're going to keep it focused this morning as best as possible, and talk first about the austere environment of space. We'll get into the mental and the emotional components that come with living in an environment beyond Earth, and, finally, we'll close with some day-to-day thoughts on what it would be like to live in space.

One of the things that inevitably happens when you have these conversations is that we get really excited about the novelty. As kids, the idea of traveling or living in space is sexy and exciting, but we don't think about two things: one, the practicality of it, but more importantly we don't think about the deeper human element of why we do it. So I thought it would be appropriate to start with a quote from Haym Benaroya, who is here with us. Dr. Benaroya has obviously spent his life thinking about living in an austere environment and building structures there. I love this quote and I just want to share it as a place to start.

-155-

> "A love of the possibilities of space exploration and settlement, initially on the Moon, but eventually on Mars and into the Solar System, is born of the positive vision it engenders for humanity. Space opens up a universe of possibilities. It creates new options for how people will view their lives and those of their descendants, but, equally important and hopeful, space also opens up avenues for improving life here on Earth and for all of its inhabitants. The engineering of space travel and lunar habitats creates technologies that can improve life here on Earth by learning how to create safe and happy environments for humans in the unforgiving space environment. We also increase our understanding of helpful and productive lives here on Earth."

I think that's a really cool way to frame the importance of this conversation. This is not just about the esoteric world of space but about our lives here. Obviously, we have some people here who are very accomplished, so I'm going to do, hopefully, the least amount of talking this morning. That said, my passion was mentioned. I was privileged to come to know SFI a few years ago. I went through their Complexity Science Summer School course, and I became entranced with what SFI is doing in the realm of complexity and in bringing extraordinary people together. But I've spent my life in the world of human performance and was privileged to serve in special operations, where we're constantly thinking about how to succeed in and train for austere environments.

We'll move through some introductions here and then jump in, and maybe we'll start with Nina Lanza, who's just up the road at the laboratory but has had an amazing life herself thinking about Mars and life beyond Earth. So, Nina, why don't you go ahead and open this up with a brief introduction?

**NINA LANZA** Sure. So I am a planetary scientist at Los Alamos National Laboratory, and what that basically means is I study rocks on other planets, specifically Mars. I'm working on the Curiosity rover mission on an instrument called ChemCam that is operating right now on Mars as we speak. And you may have actually heard that our team came up with some really big results yesterday from Mars. We found—

**B. FERGUSON** Real quick, let me interrupt you. How many people saw the news yesterday? Awesome. So it's almost like they planted a story in time for this festival. Okay, go ahead and explain what happened.

**N. LANZA** Well, amongst other things, one of the biggest, most exciting discoveries that we've made is that there is methane currently in the atmosphere on Mars. And it's not just there constantly, but actually there are little puffs of methane that appear to be seasonal, and there are several reasons why that's really exciting.

One reason is that methane on Earth only comes from two primary sources: it comes from volcanoes, but it also comes from life. We, and cows, make a lot of methane. That's a big reason we have methane in our atmosphere, but methane does not last very long. It lasts in the order of a hundred years or so, and then it breaks down. So when we see methane in the atmosphere of Mars, we know that it's being made now.

Now, we don't know what the source of this methane is, but we know something on Mars is making it right now. And there are two options for its source. It could be that there is some kind of chemistry that is making methane come out through cracks seasonally. We know of such chemistries that can do this.

But it also opens up the possibility of there being microbes on Mars. Now, don't say you heard it from me—that there's life on Mars, okay? We're still working that out, but it's a really important observation for us to track down because of its implications.

At the very least, Mars is a lot more active than we thought it was. So stay tuned, because we're still driving around Mars. We're doing it right now! So there will be many more results. We're going to try to track this methane down—that's my day job.

-157-

But the reason I'm here is because I've actually spent time in Antarctica searching for meteorites on the ice as part of the Antarctic Search for Meteorites project. We spent two months in Antarctica living in tents on the ice, equidistant between Pole Station and McMurdo, trying to find meteorites. And it's a very similar experience to long-duration space flight, so much so that NASA funds a human factor study on all of us in the field to try to figure out how we tick, what makes teams work, and what makes teams not work.

So I've had the pleasure of spending time in the field learning about what it's like to work in these extreme environments, and I think it was very surprising. So we'll talk more about that in a bit.

**B. FERGUSON** Awesome. Thank you. And I love Nina's bio, for any of you who haven't read it. I think if you were to ask me as a kid, or you asked some young kids today who were excited about space, "What is it you want to do?" They'd say something like, "I want to work on a spaceship, and I want to use lasers and go to Mars." Nina's bio says, "I'm currently living my dream, working on a spaceship, using lasers on Mars." That's pretty badass.

**N. LANZA** Greatest job ever. Right?

**B. FERGUSON** So I want to go down—we'll skip to the end here over to Ashton. Why don't you go ahead and talk about your background?

**ASHTON EATON** Sure. So I retired from the sport of track and field about two years ago. Essentially, I got into university on a sports scholarship to do this thing called the decathlon. Getting into college was my first major goal in life because my family had really never been. So being the first one was important, and once I got there I became fascinated with this challenge of physical performance because my guiding question my whole life has been, "What am I capable of?" Decathlon really let me push the limits to try to figure that out. So, essentially, for a decade I pursued that as far as I could. I went to a couple of Olympics and did the whole world record thing.

But around 2014, in between the 2012 and 2016 Olympic Games, I didn't really know if going into 2016 made sense. I had won the games in 2012, but I was thinking that there are other ways to contribute. And so I started reading a ton of books. I'd always been interested in space travel; I wanted to be a pilot when I was younger. I came across a book by Nikola Tesla, and he opened up my eyes to different ways to contribute. And then I went down this rabbit hole of philosophers and scientists. I was thinking, This is what I want to do after I'm done with sports.

I was also thinking that, while space travel's important, what about terrestrially? There are certain things that need to happen here. And I just came to the conclusion that, being out among the stars, we seek greater understanding, and we stand a better chance to understand our place. Also, it's really exciting and inspiring.

So I'm here today because I listened to a podcast featuring David— he was talking to Sam Harris, who's a philosopher, and he was talking about complexity and all these different things that I had in my mind abstractly. I have these notebooks filled with systems theory, and I was reading all of these systems theory books, and I thought, "There's a place that literally just thinks and studies this stuff?!"

So I called David and asked, "Who are you and what are you doing?" We had a conversation, and he said, "There's this InterPlanetary conference about living in space." And so I said, "I have to be there." I'll be able to talk maybe about the human performance side of things and the mentality, too, and a little bit of the physicality it takes to push your body to the limit. I believe I did it fairly well for the last decade.

..........................................................................

## Being out among the stars, we seek greater understanding, and we stand a better chance to understand our place.

..........................................................................

**B. FERGUSON** Yeah, and that's an important point. You can hear that Ashton is obviously humble in how he presents himself, but I want to contextualize this and why it's important, as we talk about living in space.

So the National Collegiate Athletic Association, the NCAA, has these statistics on the probability of becoming a professional athlete. If you play football in college, your chances to become a professional athlete are 0.3 percent. And, in the NBA, you're looking at 0.2 percent. To become an Olympian, only 0.001 percent of people who play a sport actually become an Olympian. But not only was Ashton an Olympian, he won two gold medals *and* he set a world record. And what is fascinating in that is the mindset, how do you think about it? Because on one hand there's the training, but we know there is far more on the mental side of pushing limits and training for something so specific.

When we start talking about living in space, there is a whole other level of commitment that's required. So Ashton brings a lot of interesting insight. Now we'll open up our final introduction with Haym Benaroya here. Haym, do you want to go ahead and talk about your life studies?

**HAYM BENAROYA** So it's great to be here. I'm really honored to be with three super accomplished people on the stage. My background is more along the engineering lines. I got passionate about space as a young person reading science fiction, like many people did, and watching some of these movies, and I thought, "How, as an engineer interested in structures, could I make a contribution to this adventure we call space exploration, space settlement?"

And I thought, Well, how do we build structures on a place like the Moon and make it habitable to people, so they could live happy lives? As I read more and more about it and studied it more, I started to see that, on Earth, a structural engineer doesn't have to worry about a lot of things other than building the structure—because the air is there, the sunlight is there, there are pipes for water, there's electrical conduits—whereas on the Moon, we're basically creating a whole new civilization. And so building a structure on the Moon means having people have happy lives there. If you look at this stage and you basically cut it down to about a quarter of the size that it is, that'll be the size of our first lunar structure. It's not a lot of space for maybe half a dozen people, so some of the issues that we have to deal with as engineers designing structures is also how to make that structure such that people will survive the extremes of the temperatures on the lunar surface, the radiation effects.

If you're claustrophobic, a lunar base is not for you. I know I'm a little bit claustrophobic. I don't think I could really live in a lunar base. It'd have to be a much bigger lunar base, let's put it that way.

A lot of issues that we have to deal with that are really beyond engineering, like the physiological issues, psychological issues—that's what makes it really exciting.

One of the things that we cannot test on Earth is low-gravity effects. We can do everything else on Earth as far as testing the structure against the radiation, the temperatures, and those kinds of things, but we can't test for the low gravity on the Moon. We have one-sixth of the Earth's gravity on the Moon, which means that if you can jump up five feet here on Earth, then you can jump thirty feet on the Moon. That's pretty extraordinary.

**B. FERGUSON** Sign me up.

**H. BENAROYA** Right! So the question is, what happens to the human body? We know a lot of negative things happen to the human body in the low-gravity environment, and we're starting to learn that, but the long-term gravity effects on the Moon mean that we have to live on the Moon long-term and start figuring things out. I think that structural engineering for space is really . . . it's everything engineering, so it's a really exciting adventure. -161-

**B. FERGUSON** Fantastic. And, for those who are interested, Haym runs the Center for Structures in Extreme Environments. Is that correct?

**H. BENAROYA** Right. It's a center of people that deal with these kinds of issues. And also, in case you're interested, I have a more detailed talk about lunar bases a bit later, so hopefully I'll see some of you there.

**B. FERGUSON** And I think later today he'll be signing his book. He's got a book out on building structures on the Moon, and he'll be signing it later today too.

**H. BENAROYA** That's right, later this afternoon.

**B. FERGUSON** Okay, let's dive in. I want to keep this organic, but maybe we'll start, I think, with what we call the austere environment. Because, again, when we think about space here on Earth, we forget about how unforgiving space itself can be, and I think maybe a good place to start is with Nina, who, as you heard her mention, actually lived in Antarctica in a very extreme environment. So maybe you could talk a little bit about what you learned and what surprised you the most.

**N. LANZA** Sure. Yes, so Antarctica, as you may be aware, is a continent that's on the South Pole. It's entirely glaciated, it's covered in ice, but it's the driest place in the world, so it's a desert. It's very cold there, it turns out. I can verify this. But one of the things that's so special about the project that I was on is, we are living in tents, literally canvas tents, on the ice. And there are only eight people there, and they're the only eight people you're going to see for six weeks. Antarctica is a strange place because nothing lives there. We were the only living creatures. Okay, there are penguins, but those are on the margins. Forget about *March of the Penguins*—they are mostly not in the interior. The interior is so stark. There are no plants, there are no animals, and there is no sound but the wind and the cracking of ice, and that over time becomes so alien and it makes you feel incredibly small. You realize that Antarctica doesn't care about you.

A lot of people have died in Antarctica. We think, "Oh, we have all this technology. It's easy now." And the great explorers a hundred years ago—Shackleton, Scott, Amundsen—they thought that we would have conquered Antarctica by now. They had this vision that there would be daily flights to the pole and people would live there. That's not true. It's still Shackleton's country. It's still desolate and barren and a hard place to live even with all of our technology. So one of the greatest surprises for me was that I didn't realize how beautiful our Earth is. We get so much stimulation just being here. Look around! The sky is blue, the trees are green. I hear all of these people. There's a very gentle breeze. None of that is in Antarctica. And I realized that I started feeling starved for things like smell.

There is no smell in Antarctica. You try to smell. It's very fresh air, but you can't smell anything. We're so used to being able to smell different things. Even ourselves, right? But you can barely smell anything. I also really missed darkness. I was there during the austral summer. The sun never set. It was so surprising how much I missed that. The sun kind of got lower. The sky bounced down, so you had longer and shorter shadows. But darkness is very comforting in many ways, and I really missed that.

**B. FERGUSON** Can you talk about the loneliness element, how you guys overcame that as a team? I think that's really compelling.

## There is no smell in Antarctica.

**N. LANZA** Yeah, sure. So you are really isolated. You're with a team of people that you're not related to and that you may not have met before. I had only met one of the other people previously. And so here we are living in eight-foot-by-eight-foot tents that you share with somebody you've just met. You get to know each other really well. One thing we did every evening after we came back from our fieldwork was we would gather in our communal tent, which was also only eight feet by eight feet. So we were very close together, and we would read to each other; we would read from the journals of Scott and Amundsen. Those are the two great polar explorers who eventually got to the South Pole. Of course, Scott got there five weeks after Amundsen and then didn't make it back. So we can read day by day, on the day that we are in Antarctica, we would read from both of their journals. Where were they today? What were they thinking? What were they feeling?

And it was so remarkable to hear what felt like my feelings on the page. And not only that, to share that with the team. This was our opportunity to speak to one another. We did it every night. It didn't matter how tired we were, we all gathered together and connected because we were all of the living creatures out there and we had to support each other. We literally needed each other to live.

**B. FERGUSON** And that's powerful. We were talking earlier about, if we get together here and go to space, how do we stay inspired and motivated in an extreme environment? How do we draw energy? And that ends up coming from community and rooting it into our senses. I want to go back to Ashton. Ashton made a really powerful comment about his Olympic quest and connection to senses for inspiration. Do you mind sharing that with the audience?

**A. EATON** So in 2015, it was the World Championships, a year before the Olympic Games, and the World Championships were in China. I had never been to China before, and I was there for a week or two beforehand getting prepared for the competition. It was extremely foreign to me. The food wasn't . . . the best. I was just living in a hotel, and I was feeling completely down, I didn't want to be there. It was the first time during my career where I didn't want to be somewhere.

On the morning of the competition I went out of the hotel to do my warm-up, and there were pine trees. Where I grew up, in Central Oregon, it's just full of pine trees. I went up to a tree and I smelled its bark, and instantly my entire mood was changed. I actually took a piece of that bark and put it in my pocket and kept it with me during the competition. There was just something about the mental trigger. It reminded me of why I was doing this.

First, it reminded me of where I came from, and then it reminded me why I was doing this. I think it was for home, to inspire the people where I grew up. I ended up breaking the world record at that competition because of that. And I know for a fact that it wouldn't have been the case otherwise.

So that lesson just taught me how powerful it was. And actually, I worked with Nike, leading in to the Olympics. I told them this story, and they created these vials full of all these things from my hometown. They filled one with pine tree needles, one with juniper oil, one with lava rock, so that I could grab and feel semblances of home when I'm in a foreign land.

**B. FERGUSON** So if anyone's looking for business ideas here on space exploration, it sounds like olfactory smells and vials.

**A. EATON** And you were talking about feel, how all these things make you feel, and then how you are reacting to that.

**N. LANZA** I think we undervalue scent as something that can really change our mood and our lives. None of us likes a bad smell, we all like good smells, but we don't realize how many smells are all around us. We can smell so many things right now, and that is part of what makes us feel like we are at home. And so, as you said, your town smells a certain way and when you can smell those things, suddenly you know that you're in the right place.

–165–

But when you lack that, if you were either in China or in a space station, that actually has a huge effect on your ability to do your work.

**A. EATON** Right. Well, actually one question I have for you is, if you were to go back now, how much longer could you stay, knowing what to expect? You were there for eight weeks, right?

**N. LANZA** Oh goodness. Yeah, I think I could do a lot better. I wouldn't want to go for an entire year. Those people are hardcore; I'm not ready for that yet. But if I were to go and do it again, if I were going to go longer with a bunch of scent vials and the sounds of the Earth, like rainstorms, I could definitely go for many more weeks. I bet I could do three months and still feel human.

**A. EATON** I think this is an important point. Once we've experienced something novel, I think when we go back to do it again we have that expectation. There's a preparation phase where you say, "Okay, here are the things I did wrong and right that contribute to success." Then you may be able to extend beyond that. So to your point of living in space for a long time—

**H. BENAROYA** And that's one of the challenges of going to a lunar base for the first time. We have really no such experience. And the other thing is, how do we team the people? Even though the astronauts won't all know each other, they don't know each other in that environment, and so their perceptions and how they react with each other will change with time. That's a real challenge.

**B. FERGUSON** I really want to dive into teaming in the right way, but before we do, Haym, I'd be curious to hear from you what you think [is] the biggest challenge most people don't consider when talking about living in space for an extended period of time. What is it, as someone who's studied building structures to protect from that environment?

**H. BENAROYA** So I think people don't picture the very close quarters I mentioned earlier. The other thing is that there is no outside, really, in the sense that you can go out for a walk. I mean, you can go outside with a space suit, but you're still in something.

You don't have all these colors on the lunar surface. You don't have any noises. However, on the inside of the structure, there are a lot of noises. There are machine noises, low-frequency rumbles; there are all kinds of things happening that affect your hearing.

So I think that people don't realize that you really are in a very small capsule, really forever, until eventually we can talk about larger bases. We're talking about inflatable structures that can be huge, but we're not talking about that initially. We're talking about ten or twenty years in these very small living quarters until we're able to build an infrastructure and have larger bases there.

**B. FERGUSON** You said it earlier, but could you say one more time what size space we'd be talking about? The physical space?

**H. BENAROYA** So if you want to imagine it, take one of those tents, and that would be the first lunar base—the area under one of those tents *[points to one of the shade tents, which are twenty feet by twenty feet]*—and that includes people with beds, it includes machines, it includes places where you eat, places where you shower, and that's maybe for half a dozen people.

**B. FERGUSON** So all the recruits will come from San Francisco? It's that kind of experience.

**A. EATON** I live in San Francisco now, it's basically the same thing.

**N. LANZA** I remember the principal investigator of the Antarctic project that I'm on actually put it really well. He said, "You know, this experience of living in this small space is something that most of us don't ever do, except with a spouse-like partner." We share with maybe one-ish person. One person can have that space, but you're going to have to do that with more people, and people that you're not married to or otherwise partnered with, maybe people that you don't like all that much. Hopefully, the first settlers will be a really tight team. As we start building, you're not going to love everybody who lives in your lunar settlement. So how do you deal with that in these quarters, the size of that tent?

-167-

**B. FERGUSON** Unfortunately, we're down to a few minutes, so let's dive into teams. I think this is something really compelling, particularly given the experience on the stage.

I will say that when we leave, grab these guys throughout the day. There's a whole bunch we didn't even get to touch on. One thing that I find is fascinating that everyone's been involved in is building equipment for extreme environments. Ashton is coming at it from someone who designed tools and equipment for extreme sports at Nike, to amplify an athlete. And then we've got Nina working on the Curiosity rover, and Haym, who has spent a career thinking about structures. It's a fascinating conversation but, on the teams front, back to you, Ashton. When you think about it, if we're going to live in a tent that size for any extended period of time, you can't walk away if you have a falling out with one of your teammates. So, coming at it as an Olympian on high-performing teams, what is it that you think matters most?

**A. EATON** Communication. This is something I actually didn't do very well at all; my coach and my wife taught me. My wife is also an Olympian, and we trained together, and I think the source of a lot of, maybe all, conflict, at its very base level, is misunderstanding. So being able to say, "Haym, I understand or know to a certain degree where you were coming from." And for Haym to say, "Ashton, I know where you were coming from," to have some sort of mutual understanding, is unbelievably massive. You actually can't articulate it.

If you have a great relationship with somebody, it's most likely at the basis communication. I want to touch on something outside of that—and I think Nina was kind of bringing it up—when you are going to these new environments, the information that you bring back with you is absolutely vital. I don't know, you probably journaled, but I think there has to be gear or a component to the things that we make that captures the information about living in that environment, because there will be a lot of unconscious and hard-to-articulate things in our mind. So if we can somehow have the environment help us gather that data and bring it back, I think that's very vital to the next group.

**B. FERGUSON** Go ahead.

**N. LANZA** Well, I think that is really what we're going to need to do. It's very hard to predict how you're going to feel, and the only way to figure that out is to put yourself into these extreme environments. We can theorize all we want, but we are going to have to live in extreme environments in order to learn how to live in extreme environments.

**B. FERGUSON** On the team front, what was your most profound insight, even though it was for a shorter period of time, from living with the team in Antarctica?

**N. LANZA** Well, I think I have to agree. I mean, this is what everyone says—communication—but that's absolutely the case. My tentmate really needed her space in the morning, but eight feet by eight feet is not a lot of space.

And so she said, "Look, let's not talk in the morning." She had headphones in, she was listening to some music, she had her hoodie up so she couldn't even see me. It's not that she hates me—we actually are great friends, we are wonderful friends—but she needed her space. She needed to feel like she was alone in the morning. And you can't be alone in Antarctica. It's not safe. So she expressed this need to me, and I was really glad she told me. I think being able to ask for what you need honestly and being able to accept it and help other people meet their needs—that give and take is key. It's so important because, I mean seriously, this is why people kill each other in these tiny spaces, right? If their needs aren't met and your needs aren't met, this is an unhappy team and you're not going to be able to accomplish what you're both there to do.

> ### This is why people kill each other in these tiny spaces.

**B. FERGUSON** Right! So Haym, when do you think we might actually have a livable structure beyond Earth, if you had to give an over-under on a date?

**H. BENAROYA** So I would say that politics governs everything, it seems. I would say if the politics are right, we could really be on the Moon with a capsule in about ten years. And then once we're there, then I think within ten years after that, we can have much larger structures, and from there we can build the infrastructure and understanding we would need to be able to go to Mars and perhaps to do other things that we're talking about that are very exciting.

**B. FERGUSON** Ten years—you heard it here first. Who's interested in going, show of hands? Not even half! All right, we can talk after. I want to thank you guys for your time. I recognize we just scratched the surface. I think everyone here will be around all day. There are a lot of compelling questions and conversations we want to get to. But it's been a privilege, and I'm humbled to be up here with all of you. So thank you guys. Thanks to SFI.

**A. EATON** Thanks for coming out.

**N. LANZA** Thanks!

# INTRODUCTION:
# THE ORIGINS OF LIFE IN SPACE

In 1961, Frank Drake famously created an equation for thinking about the likelihood of detecting intelligent life beyond our own planet. This equation multiplies together the rate of star formation, the probabilities of life arising and persisting, the number of inhabitable planets per star, and the duration of the emission of detectable signals from another civilization. While recent advances have allowed the scientific community to determine many of these numbers (we now know that nearly all stars have at least one planet) and as debate continues about detectable signals (there are serious questions about whether radio waves make it very far in space), we still do not have a grasp on one of the most fundamental probabilities: How likely is it for life to arise in the first place? This implies that much of astrobiology is effectively concerned with the same questions as one of the great mysteries on our own planet, the origin of life. How likely is it for life to begin? What conditions are necessary? How likely is it to persist? How likely is it to evolve greater complexity?

It is important to recognize why these questions are so challenging. For understanding how life arose, we are trying to constrain a process that we cannot yet create in the lab: the transition from complex chemistry to truly living matter. Modern life gives us some insights into the past history of life, but the tricky part is that even the simplest cells are amazingly complex, with a large number of chemical processes that are precisely orchestrated to dynamically respond to environments and eventually reproduce. At the simplest end of the spectrum, we understand how to make simple chemical replicators—known as autocatalytic sets or cycles—which give rise to more of their own chemical elements. What we lack is the ability to take basic replicating chemistry and to marry it with an evolutionary process that evolves greater complexity and myriad biological outcomes. This idea, which is often called *open-ended evolution*, is really a statement about how adaptable a system is, and we believe that it is a critical feature for life's ability to persist. One of the great challenges in understanding how an evolving chemical system eventually becomes life is that we don't know the rarity of the "right" environment, and we don't know how long this environment needs to remain roughly undisturbed to evolve greater complexity. We might find some hope in the best estimates for the timescale to evolve the first cells, which is roughly 500 million years and relatively quick in geologic terms (for example, it took another 1,500 to 2,000 million years to evolve single-cell eukaryotes, which are only slightly more complex).

> One of the great challenges in understanding how an evolving chemical system eventually becomes life is that we don't know the rarity of the "right" environment, and we don't know how long this environment needs to remain roughly undisturbed to evolve greater complexity.

Another great challenge for both the origins of life and astrobiology is determining how different life could be. Extant life has and continues to provide us with countless surprises in terms of its diversity, functionality, and existence in unlikely environments. Classic macroscopic biology was continually surprised by the scope of biological diversity (e.g., carnivorous plants), and modern microbiology has been even more surprising where single-cell organisms have been found living in environments of extreme temperature, acidity, pressure, and salinity. These recent microbial discoveries have greatly expanded our conception of life's physiology and metabolism, perhaps giving us hope that life could be common in the Universe, having seen its apparent ability to adapt and survive on Earth.

All of these considerations for the probabilities of life beginning—and the great diversity that it may ultimately come to display—should be seen in the context of a variety of near-term methods for potentially detecting life. For more than fifty years, the scientific community has conceived of ways to look for life in our own Solar System. These began with questions about whether the Viking lander would see macroscopic life on Mars in 1976, and have been refined to considerations of potential microbial life in subsurface deposits of frozen water on Mars, evidence of extinct life on Mars, or searches for life in the most-likely liquid subsurface of the frozen Saturnian moon Enceladus. There are a variety of possible future NASA probe missions to search out these environments with the real possibility of discovering life. However, other recent advances provide us with the possibility of searching not only a few worlds within our own Solar System but thousands of worlds across the galaxy. These advances stem from the discovery of exoplanets, planets around stars that are not the Sun. We have gone from questions about whether planets exist around other stars to the initial verification of an exoplanet in 1992 to a catalog of thousands of exoplanets and the realization that almost all stars have at least one planet. Even more exciting is the burgeoning technique of determining what is in the atmospheres of these planets by looking at the spectrum of light that has come from the star, passed through the atmosphere of the exoplanet, and made its way to our telescopes. Detailed atmospheric composition will come from newly launched and future space telescopes, and determining whether life is responsible for this composition again becomes a complicated question of what type of life is there and what metabolism it has. How complex is this life? What fraction of the planet is occupied by life? What is the diversity and ecology of life on a planet? How do all of these features combine to regulate the dynamics of an atmospheric composition, giving distinct biosignatures?

Each of these techniques for discovering life is ultimately limited by hard questions about how different life could be. To address this, we need new theories for defining the broad possibilities of life. We need to generalize our current theories of biology by connecting them with fundamental physical and chemical laws. For example, the laws of thermodynamics generally constrain the possibility for the bioenergetics of metabolism, and we understand how to abstract the process of evolution to a variety of contexts and systems. What we need are detailed theories that integrate specific and general knowledge of biophysics, evolution, biological morphology, physiology, metabolism, genetics, and information storage. Such theories should be useful for interpreting our own evolutionary history and for predicting modern ecology. They should be useful for predicting a range of biological physiology and potential ecologies. They should be able to scale up to planetary-scale interactions and regulation.

The following conversation touches on many of the issues related to the likelihood of life beginning on a planet—as opposed to being transported to a planet where it might thrive—where the focus is on the vast space of chemical possibilities and the natural capacity of complex physical systems to give rise to emergent patterns. We also discuss the difficulty of looking for signatures of life that could be different from those on our own planet, the need for general principles in looking for signs of life, and the promise of finding signs of extant or extinct life on various planetary bodies in our own Solar System. We also describe the excitement around observing the properties of exoplanets and the need to develop theories for what a living planet looks like.

*—Chris Kempes*
*Professor, Santa Fe Institute*

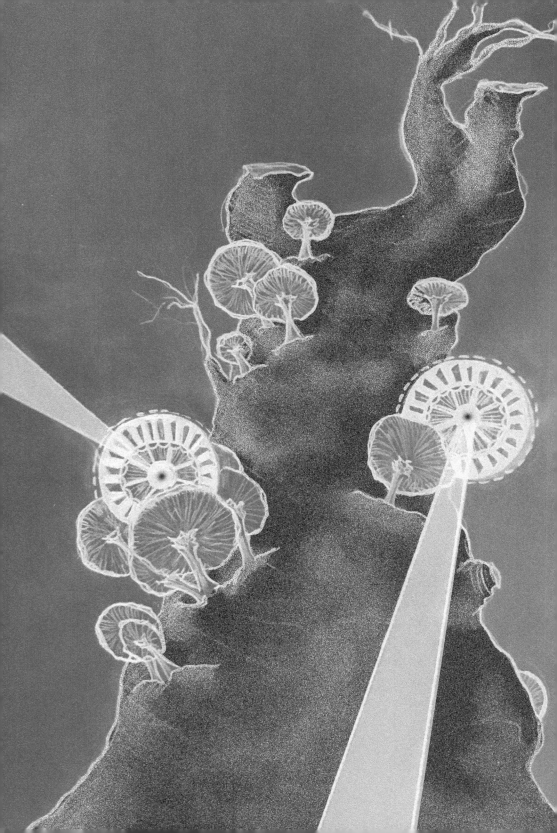

# THE ORIGINS OF LIFE IN SPACE

*D.A. Wallach introduces the panel,*
*moderated by Chris Kempes and featuring Caleb Scharf.*

**D.A. WALLACH** We are now going to witness a discussion about the origins of life, and we couldn't have better participants. Chris Kempes, currently a professor at the Santa Fe Institute, is an astrobiologist and physicist—very interested in the origins of life. And Chris, prior to SFI, spent time at NASA Ames and Caltech. He also has a PhD in physical biology from MIT.

Across from him, Caleb Scharf joins us from New York City, where he is at Columbia University. And also in New York, he is the cofounder of YHouse, a new institute devoted to uncovering the mysteries of consciousness. But when he's not focused on consciousness, he focuses on astrobiology, physics, and the origins of life as well. So with that, I will turn it over to them. You are in for a treat!

**CALEB SCHARF** Thank you.

**CHRIS KEMPES** Yeah, thank you, D.A.

APPLAUSE

**C. SCHARF** So we're going to talk about the origins of life in space. But what does that mean?

**C. KEMPES** Yeah, so, by "life in space" we don't mean the life that we put in space. We mean homegrown, naturally occurring, free-range life that has self-organized itself on some planet. This is a question that's intimately connected with astrobiology. In order to understand where there might be life in the Universe, we have to first understand how life gets started.

So, in thinking about astrobiology, one of the main organizing principles for the last forty years has been the Drake Equation, which is just a way to put together all of the different probabilities that go into the likelihood of finding life in the Universe. And so I wanted to ask you, Caleb, what do you think are the hardest parts or most unlikely probabilities in that Drake Equation? Which things do you think we understand really well, and where is there room to discover more?

**C. SCHARF** Sure. And just to add to this topic that we're talking about, I think part of the motivation—because sometimes people say, "Well, why do we care about life elsewhere? Sure, it would be cool to know"—but, ultimately, it's about finding context for ourselves. And the Drake Equation was sort of at the forefront of that. Frank Drake, an astronomer, came up with this back in 1961. I'll just explain that a little bit, and then we'll talk about the pieces we know something about, as well as the pieces we are entirely ignorant about, but where we're hoping to make progress.

So the idea of the Drake Equation is really quite a specific thing. It's about estimating how many communicative species are out there in our galaxy. Now it's not a first-principles equation. And it's not a predictive equation, because we have no idea how to write an equation like that, which I think is something we'll talk a little bit more about. The Drake Equation is a set of factors that, if you multiply them all together, gives an estimate for how many communicative species there could be at this moment in our galaxy.

So how do you do that? Well, the first set of factors in this equation is things that we actually have some knowledge about. The very first factor is, how many stars are formed in a given amount of time in our galaxy? So how many new stars are formed or come into existence every year? It turns out that it's in the order of one to two stars. That's something that astronomers have been able to pin down.

## DRAKE EQUATION

$$N = R^* \cdot f_p \cdot n_e \cdot f_l \cdot f_i \cdot f_c \cdot L, \text{ where}$$

$N$ = the number of civilizations in our galaxy with which communication might be possible;

$R^*$ = the average rate of star formation per year in our galaxy;

$f_p$ = the fraction of those stars that have planets;

$n_e$ = the average number of planets that can potentially support life per star that has planets;

$f_l$ = the fraction of the above that actually go on to develop life at some point;

$f_i$ = the fraction of the above that actually go on to develop intelligent life;

$f_c$ = the fraction of civilizations that develop a technology that releases detectable signs of their existence into space; and

$L$ = the length of time for which such civilizations release detectable signals into space.

Next, we multiply that by a fraction or a frequency that says, well, how many of those stars build planetary systems around them? Because planets, we think, are important environments to look at in terms of life. And that, too, is actually a number, a quantity, we now have some constraints on. I think if you were here yesterday, you heard some discussion of the work being done on so-called exoplanets. And we'll talk more about that as well, as we have some idea. The bottom line is that every star in our galaxy is likely to have some kind of planetary system about it. We actually have good enough statistics to make that statement.

And the next factor is, well, how many of those factors might resemble the Earth in some way? They might be smallish, rocky planets that could be temperate. And even there we have some inkling of what that number is; it's somewhere between ten and forty percent of planets, we think. That's only saying that these planets might be the same size as the Earth, possibly a similar composition. But then the next factor is, what is the frequency of abiogenesis? How often does life occur on a seemingly suitable planet? And that's where we hit a brick wall. That's where our knowledge comes to a grinding halt, because we actually have a really hard time saying anything about the frequency with which we expect life to occur in these environments.

I know this is something we want to get a little bit deeper into—these questions of origins of life. And maybe one piece of that is just defining *life*. Chris, I think it is true to say that it sounds so simple: *What's life and what's not life?* But, actually, it's a really difficult question.

**C. KEMPES** Yeah, I agree. There are long debates about how to define life, and I don't want to get too much into that, because there are a lot of details that people really care about. For me, what it mostly comes down to is things that are able to replicate, persist, and evolve greater complexity. So much of what we require of life on our own planet is for it to be able to find new niches, new environments, be able to basically spread, avoid extinction, find greater functionality, and adapt to particular types of environments.

But I also want to say that I really agree with you, Caleb, about how hard estimating this probability of the origin of life itself actually is. Most of the time, when we want to calculate a probability, we have this massive set of data and we take some nice average over it. Even in the exoplanet world, we now have enough different types of sets of information that we can calculate really nice averages. But for the origins of life, we have only this one event on our own planet, and so it's hard to extrapolate that to think about how likely that is in general. And so a lot of the work then needs to focus on the principles, that lead to something like an origin event. And from there, you can start to ask, well, where are those conditions likely to be somewhere else?

But that also remains a very hard problem, because the space of chemistry is both large and complex, and the types of things that we've been able to see in different complicated environments gives us hope that forming different types of complexities is perhaps easy. And a lot of the focus in this world of chemical origins is on autocatalytic cycles, which are just cycles that give rise to more of themselves. That starts to look like replication, in a very crude sense—where you have some cycle that's propagating the cycle into adjacent environments. A lot of the evolutionary theory that we see today can even be used to think about how different types of chemical systems compete with one another, how those might be selected for in time, and perhaps how they might evolve greater complexity.

We also know that lots of different types of physical structures can be organized simply from physics. So there are lots of interesting patterns that you can get completely without life, just with simple chemical and physical systems. You may have some chemical that is diffusing in an environment—that's just a physical process of how random spread occurs—and then also reacting in an interesting way, such that these emergent patterns happen. And so this tells us that there are lots of interesting processes that can occur before life that look sort of similar to life in lots of ways.

So that's all good news for starting to uncover some of the processes that might allow us to calculate how likely the origination of life is. If we then look at our particular history to try and get a better sense of that probability, there are places where we can be really optimistic and places where we can be really pessimistic.

The optimism is that it looks like it took us about half a billion years to go from no life to something like bacteria after the conditions were right on this planet. So after something like a moon-forming impact, which would have sterilized any life that had existed at that point, we then don't wait very long, only half a billion years, to see a new form of life, or the first form of life. That makes us fairly optimistic. But what makes us pessimistic is that we then take at least a billion years to go from this very simple form of life to something slightly more complicated in the eukaryote. It takes much longer than getting life in the first place. So the fact that we go from existing life to slightly more complicated life in a much longer timescale might provide some pessimism in terms of how easy some of these steps are. I think this is still a probability that we're uncertain about. But I'm wondering, Caleb, what would change your estimation one way or another about how likely that event is?

> The optimism is that it looks like it took us about half a billion years to go from no life to something like bacteria after the conditions were right on this planet . . . . But what makes us pessimistic is that we then take at least a billion years to go from this very simple form of life to something *slightly* more complicated in the eukaryote. It takes much longer than getting life in the first place.

**C. SCHARF** Well, I think what's interesting, listening to you describe some of your insights to the processes that we think must be involved in giving rise to life as we know it—and, again, we still don't have a perfect definition of what's living and what's not living—but one thing that occurs to me is, again, if in the Drake Equation, that magic expression where we have the probability of this environment, this planet giving rise to life, all the pieces that you describe have yet to be pulled together into a fundamental theory of life. Now, I think that's one of the biggest open questions in science right now. We do not have a first principles, fundamental theory of life in the way that we have theories in physics and so on, and that is one root to calculating that probability, to answering the question of whether we're alone or not, but then also reflecting on that in terms of our existence here. But there's another way to do it, right? I mean, there's the first-principles calculation that says, this is the probability—and, incidentally, I think it's also fair to say we have no idea whether that's a reasonably high probability or an extremely tiny probability for a given planet. That's another question to look into.

We don't have the first-principles theory to give us an estimate of that probability, but we have another route, which is actually to forget about the details and to go looking for life in the Universe. And by looking for life in the Universe, if we could count other instances of life across our galaxy, we could then derive from that a constraint on that fundamental probability. Now, you probably need both of these things to really understand the nature of life, but it's kind of interesting to me that I don't know which is going to succeed first, and they're going to ultimately help constrain each other. I mean, maybe, there's a question of just how unlikely life is, it is an interesting one. And it always comes up. Do you have any feelings on that?

**C. KEMPES** I don't. I wouldn't be able to put a number to it. I would say my estimation would radically change if we see another, even bacterial, species somewhere in our own Solar System. And that's where a lot of the hunt is happening, beyond our own planets. We've heard a bit about what's happening on Mars and that search earlier today, and then there's also a search on places like Enceladus, which is an icy moon. And so I think if we saw something else there, obviously that for me makes the probability one hundred percent.

For a Solar System level, it would probably put me more at that end of the scale.

**C. SCHARF** Right, yeah. And I think . . . well, there's some interesting minutiae to this as well, because one way to have the probability of life on a planet be quite high is to have the planet be a complex chemical incubator. A planet is a pretty big place, and the probability of a particular set of molecules spontaneously ending up in the right place and the right configuration to give you that first autocatalytic cycle or whatever may be extraordinarily small. The scale of a planet is like a Google system. It's searching this enormous chemical landscape, but it's doing so many simultaneous searches. It's like a big parallel computation. Do you feel that that plays into it? I mean, is it more probable to find life on a larger planet than it is on a small, icy moon? And then that's related to the question of the persistence of life.

**C. KEMPES** Yeah, I think for me it's more a question about the ability for life to move beyond the place where it starts into other environments. So I think that a small planet that's incredibly rich in its chemistry and energy flux, where you can get life to start in one place and then propagate across an entire planet, that to me makes persistence and evolutionary complexity much more possible. That to me is the real question. So if you have an icy moon, it has one really great vent system where there's huge amounts of energy production, it's even possible that in such a great environment life starts, but for reasons of circumstance and happenstance it just doesn't happen to make it, because it can't propagate into nearby adjacent environments. I think that's really the question: How rich is a planet? And, also, how easy is it for life to start and start to give rise to these feedbacks that make its persistence more likely?

**C. SCHARF** Right.

**C. KEMPES** Actually, on that topic, we've talked a little bit about searching for life in our Solar System, but I want to switch gears to think about one of the best places to search for life, which is across many solar systems. This is something from a "go there and look" perspective that we didn't really think we'd ever be able to do, but now with these new observational techniques that are allowing us to see exoplanetary environments, where we can see planets going around other stars, and even start to look in the atmosphere of those planets, we now have a much greater sample of environments in which we might be able to look for what we call "biosignatures," or signatures of life.

More new technology is coming online to give us better resolution of what's in the atmospheres of these other planets. This is something Caleb does a lot of, and so I'm wondering, what do you think is the timescale for getting the type of resolution we need so that we might be able to accurately see a biosignature in another planet?

**C. SCHARF** Yeah, that's a great question. So just to try to summarize it, as you say, we've developed these various techniques for detecting . . .

TRAIN PASSES BEHIND STAGE

**C. SCHARF** . . . the presence of planets around other stars, or the presence of noise—not trains but other things. Some of those techniques, in fact the majority of those techniques, rely on the effect of the planet on the star. Either the gravitational influence of the planet makes the star wobble a little bit around its location or one of the most successful techniques, the so-called transit technique, is essentially a miniature eclipse. So just by chance, in some exoplanetary systems, the orbits are aligned so that from our point of view the planet moves between us and the star and diminishes a tiny bit of the light of the star when that happens. It happens once every orbit. So we can use that to deduce the existence of a planet, the orbital period of the planet, and hence how far the planet is from the star and a constraint on the size of the planet.

Now, something else we can observe with transits is when the planet is between us and the parent star, if the planet has atmosphere, some of the starlight will be filtered through that atmosphere. And by looking at that filtered light, by splitting it into a spectrum, we can actually, in principle, detect the molecules and atoms that make up the atmosphere. And this is one of the prime techniques that we hope to deploy in order to look for signs of so-called biomarkers. On Earth, oxygen is a pretty good biomarker; so is something like methane. These are things that are predominantly produced by biology.

So we're looking for these atmospheric chemical signatures. Now, when is that going to succeed in terms of looking for life? Well, right now we have a few non-Earth-like planets where this has been done. And giant planets, more like Jupiter, also called super-Earth planets, fall into two classes, one of which may be more Earth analog than not. The reason we've been able to make some of those chemical measurements on those planets is because they're bigger, and so there's more atmosphere and there is a greater chance of getting a decent signal using a telescope. What's going to happen in the next few years, we hope in 2020, is that NASA is going to launch its next really big telescope, the James Webb Space Telescope, called JWST. I won't say how much it costs [*whispers into mic*] nine billion dollars—

LAUGHTER

**C. SCHARF** —due to various cost overruns. It's an extremely ambitious instrument, though, and if it all works, it will be a remarkable instrument. So that's supposed to go up in 2020. It may be within a year or two of that launch that we begin to get some of these measurements of potential Earth-analog planets orbiting other stars, where we can look at the spectra of light filter through the atmosphere and see what the composition is, or put some constraints on the composition.

Beyond that, I think within a decade we ought to be able to say something about these potential life-harboring, Earth-analog planets, whether any of them show signs or could be consistent with a biosphere. We have discovered as well—and maybe you were going to talk about this—that there are things that will trick you. We used to think oxygen—I mentioned that oxygen and methane were perfect biosignatures. The only reason they exist in an atmosphere is something's continually putting them into an atmosphere, and on Earth we know methane and oxygen come from life, predominantly.

Well, it turns out that that may not be true. In fact, there are non-biological processes that can enrich an atmosphere with oxygen depending on the chemistry of the rock surface of the planet, and even things like methane become very difficult. So it's going to be interesting, and it's actually going to start happening pretty soon, in the next few years.

> Within a decade we ought to be able to say something about these potential life-harboring, Earth-analog planets, whether any of them shows signs or could be consistent with a biosphere.

**C. KEMPES** Yeah, and I think this point about how hard it is to pick a signal that you can prove is not produced by some abiotic process is really very challenging, because it comes down to building entire planetary models. And the problem when you build an entire planetary model is that you have all of these different varieties of coupled dynamics. You have this very complicated atmosphere that has these large-scale and small-scale fluid dynamics. You have an ocean that's doing the same thing. You have a particular continental geometry. You have a tapping-in in the core of the planet, and all the volcanic activity, and then if you inject biology into that, you have all of these interesting feedbacks between how the ocean uptakes and releases certain types of chemicals from the atmosphere, for example, oxygen or carbon dioxide. And that's interacting with the fluid and giving rise to these planetary-scale processes.

And so when you start to couple that many different types of dynamics together, each of which has very complicated physics, very complicated chemistry, then you run into this challenge of being able to invent fairly exotic processes that are without life, that will give rise to signatures that look very lifelike. And, in a way, that's sort of what we need to get life to start in the first place. We need something that is purely abiotic that starts to give very complicated types of signatures and almost could fool us into thinking that it's lifelike; otherwise life couldn't start in the first place.

......................................................................

We may see planets that reflect the Earth as it perhaps was when it was only a billion years old, and that will be extraordinarily interesting, not just from the point of view from studying those worlds but what we learn about the potential past history of environment on our planet, and in the future.

......................................................................

I think what's interesting is that a lot of the work that's happening in astrobiology, especially around exoplanets, is intimately connected with the types of planetary ecological models that we build for our own planet. And that's a very nice connection, because it means that all of the ecological thinking that we're deploying to think about the long-term dynamics of our own planet and climate change will allow us to say things about other exoplanets. But also, all of the work that happens in this exoplanetary sphere can tell us something about what we might be missing in the dynamics of our own world. I think this is, from an interplanetary perspective, a really nice connection between the very far-away and the very close types of planetary dynamics.

**C. SCHARF** Yeah, and just a couple of things to add to that. So I'm an astrophysicist by training. The funny thing is, planets are horribly complicated. In astrophysics, stars are simple. We can figure out the basic functioning of stars with just a few lines of equations and a little bit of physics. Planets are insanely complex. It doesn't matter whether it's a gas giant like Jupiter or a terrestrial-type planet like the Earth. But the interesting thing, as Chris says, is that we're slowly realizing as we do this work—for example, I'm involved in building sophisticated supercomputer models of planetary climate—not for the Earth but for alien worlds, eight worlds in different configurations—and what we're realizing is, it's a mirror to us. Out there will be exoplanets, Earth-type planets that reflect the Earth as it has been in the past and as it will be in the future.

-187-

So we may see planets that reflect the Earth as it perhaps was when it was only a billion years old, and that will be extraordinarily interesting, not just from the point of view of studying those worlds but what we learn about the potential past history of environment on our planet, and in the future. I mean, we're always interested in the future, and I think this festival is in most respects about the future, about our future as a species, about the integrated system that we're part of. There may be worlds out there that may be Earth ten billion years in the future, around a different type of star that will live that long. Our Sun, unfortunately is not going to take us that far. In fact, in about a billion years' time, it's all going to be over for life on the surface of the Earth. But, yes, the complexity is an enormous challenge, but it's also an extraordinary opportunity to learn new things about the functioning of our own world.

**C. KEMPES** Yeah, and on that note, I would also say that a lot of what I see happen in scientific progress is either we develop some nice fundamental theories that tell us what the bounding possibilities are in some system, and we often use those to great effect. But another thing that has really pushed our progress in the science along is surprising observations, and that usually requires looking at a vast amount of diversity. And so, especially in the biological sciences, I would say that observing forms of life that we didn't think could exist, or observing types of metabolic processes in cells or communities, all told us something about where our thinking was wrong, how diverse life could be, or what evolution actually looks like. I think it's very exciting to suddenly have this opportunity to see a variety of planets so that we can get surprised and better inform our thinking about where we have it right about life and planetary dynamics and our own planet. And also where we are surprised about what else could be possible. So I think that's a place where we should be able to make a lot of progress in the next twenty to fifty years.

**C. SCHARF** Yeah, absolutely. And, to come back to the question of identifying life and biospheres and seeing whether they resemble anything that we know on Earth, and, as Chris says, life on Earth is vastly more diverse than we had imagined twenty to thirty years ago. I mean, there are crazy, crazy organisms that we're not necessarily used to, because they're in environments that we don't inhabit. There are things that have been found deep in South African gold mines, in isolated pockets of water, where actually the root feedstock for their metabolic cycle originates with radioactive decay in the rocks, which actually splits water molecules and produces useful things like hydrogen and oxygen that those organisms gobble up. And without that radioactive decay, they would not be able to survive in those environments.

But from the astrobiological perspective, from the exoplanet perspective, we may struggle to look at a single planet and say for sure that that has a biosphere, and that it is a biosphere engaged in these sorts of processes. But my suspicion is, one way this may play out, we'll see in the next ten years, is that we'll find a whole bunch of planets that just look different. That something is going on on that set of planets that isn't going on on this whole other set of planets that are otherwise very similar. And I think we'll be drawn to the conclusion that life is happening to those planets, and not to these other planets. And that will actually be incredibly informative, because then we can immediately start asking what went wrong, what went right, depending on your perspective. I mean, you could see life as the ultimate infestation, and planets may not want to be infested by life. But it definitely changes them significantly.

**C. KEMPES** Yeah, absolutely. Well, I think that's actually just about our time. So thank you, Caleb, for participating in this conversation.

-189-

**C. SCHARF** Thank you. This was fun.

**C. KEMPES** And thank you all for coming today.

# INTRODUCTION:
# INTELLIGENT SYSTEMS

Imagine arriving on the surface of an exoplanet. Consider what it would take to convince yourself that you had encountered an intelligent agent. Leaving aside indisputable proof in the shape of giant menacing robots, spaceships, and gleaming cityscapes, what clues would you seek as evidence of intelligence? Better (or worse) still, as artifacts and monuments to stupidity?

And, having succeeded, what would you do to establish your own intelligence—which single behavior might be met with over-whelming confidence in your abilities rather than alien dismay at your evident stupidity? Setting up a Turing test is ruled out; the whole language game of question and answer is predicated on a shared symbol system and grammar—precisely that which we cannot assume.

Finding evidence of life seems relatively straightforward next to identifying intelligent life. A planet surface scattered with pulsating orange sheets of lichen-like netting moving about with spooky purpose and growing to encompass the surface would do the trick. Evidence of metabolic activity and reproduction tend to satisfy most of the requirements that astrobiologists have asked of life.

Intelligence is not as straightforward. It is compounded by the challenge of discovering a means of establishing intelligence in both organically evolved life forms and in culturally engineered technologies.

For organic life, the first evidence that an agent is intelligent would probably take the form of strategic behavior: avoiding obstacles, pursuing prey, moving toward the best sources of light and energy and away from those regions with little to offer. Intelligent life would not only be captivating, it would be uncanny.

It is not nearly as obvious what an intelligent solid-state quantum computer might do, other than look menacing in its studied, inert indifference. Arthur C. Clarke and Stanley Kubrick imagined an intelligent monolith in their film *2001: A Space Odyssey*—about as dynamic as a fossilized turd. At least they gave the computer HAL an aperture over its red eye so that we might know when it was thinking.

-191-

................................................................................................

# Leaving aside indisputable proof in the shape of giant menacing robots, spaceships, and gleaming cityscapes, what clues would you seek as evidence of intelligence?

................................................................................................

An experimental approach to organic life should extend beyond strategy toward discovering evidence for sophisticated forms of inference. Learning that the rocks you offer an agent are of little interest, whereas the salted peanuts are worth the effort: avoiding the prod and approaching the trough; better still, learning a sequence of colors or sounds or numbers, and solving simple classification tasks like sorting rectilinear objects from spherical ones. These latter are of course the bread and butter of machine-learning algorithms in our solid-state devices, which many insist show no signs of intelligence, or are merely "programmed to look intelligent."

Perhaps the ultimate confirmation that we are dealing with an intelligent agent—organic or solid-state—would be the acquisition of some form of representation. In response to your queries, a lichen-like creature (or a crystal matrix, or a panel of LEDs, or an interfering pattern of mercury wave-fronts) changes shape so as to inscribe an image that more effectively represents a favorite object, or a sequence of glyphs that encode regularities inherent in the tasks that you have set it.

The representational approach to proof was pursued by Denis Villeneuve in his film *Arrival*—based on Ted Chiang's short story "Story of Your Life"—in its use of a time-symmetric, three-dimensionally immersed, self-organizing, ink language to communicate past and future events.

A simple means of summarizing many of these ideas is to observe that strategy, inference, and representation are definitive signatures of intelligent life. All life is thereby to some degree intelligent, and in equal and opposite degree, stupid. Organic and cultural evolution imply that all evolved lineages, with even a reasonable fit to their environments, are capable of strategic behavior, making meaningful distinctions through inferential choices, and encoding these behaviors in representations of logical schema or memory systems. Hence the organic and the inorganic, brain and machine, scientific and artistic, rational and motoric, and the intuited versus the educationally derived can all contribute in different ways to augmenting these essential capabilities.

With this more capacious, tripartite approach to intelligence, we might be in a better position to recognize and appreciate it elsewhere (on our own planet and beyond) rather than search endlessly for the multiplying mirrors of the human mind.

The intelligent design panel sought to explore intelligence through this very broad interplanetary perspective, including the historical neglect of nonhuman intelligence and its causes, the role of educational systems in developing or hindering intelligence, the similarities and dissimilarities between nervous systems and deep neural networks, and the multiplicity of ways that the future and diversity of intelligence has been imagined and predicted through film.

One of the insights that came up repeatedly in different forms in this panel is how dangerous parochial views of intelligence might be for the planet: in education, the obsession with IQ and accreditation rather than self-directed learning; in animal intelligence, the preoccupation with human reason and language rather than diverse intelligences; in artificial intelligence, the use of flawed and biased commercial human data sets to train machine-learning algorithms rather than more representative demographic data; and the sociopolitical implications of current technological trends on less paradigmatic but no less important intelligent capabilities like empathy, creativity, and the imagination.

*—David Krakauer*    -193-
*President and William H. Miller Professor*
*of Complex Systems, Santa Fe Institute*

# INTELLIGENT SYSTEMS

*D.A. Wallach introduces the panel,*
*moderated by David Krakauer and featuring Vanessa Ferdinand,*
*Jonah Nolan, Graham Spencer, and Kurt Squire.*

**D.A. WALLACH** How's everyone feeling? Come on, I heard Santa Fe was where the crazy people live. I want to see how crazy you all are. Let me hear a primal scream!

HOLLERS

**D.A. WALLACH** That's exactly what I wanted! Okay, this panel, I will venture to say, may be one of the finest in this very fine two-day event, and it is about intelligent systems. It's no accident that one of the most intelligent people in Santa Fe, David Krakauer, is both involved in it and sort of moderating it, although we hope it's going to be more of a free-for-all. David, I think, requires no introduction, and therefore I won't give him one.

**DAVID KRAKAUER** *Laughs.* Good, man!

**D.A. WALLACH** Vanessa Ferdinand is many things. She has been an anthropologist, she studies cognitive systems, and recently as I understand it, she studies information flow through evolutionary processes.

We also have Graham Spencer here with us. Graham, most importantly, is on the board of the Santa Fe Institute.

LAUGHTER

**D.A. WALLACH** But, secondarily, he is a venture capitalist at Google Ventures. He became so after selling a company that he had founded to Google, and that was just one of many businesses that he's been a part of building.

Jonah Nolan is a personal friend. I'm very excited to have you here, Jonah. Jonah is from Los Angeles and his discography and bibliography and . . . what is the television equivalent?

**D. KRAKAUER** Cinematography?

**JONAH NOLAN** Sure, we'll take that.

**D.A. WALLACH** His Scientology—

LAUGHTER

**D.A. WALLACH** —is extensive, but if you watch, for example, *Westworld*, or if you've seen *The Prestige* or *Memento* or *The Dark Knight* or *Interstellar*, you've been lucky enough to enjoy Jonah's work, which deals with a great diversity of scientific and theoretical topics. And finally but most importantly Kurt Squire. Everyone, give it up for Kurt Squire!

**D. KRAKAUER** Yay.

APPLAUSE

**D.A. WALLACH** Kurt comes to us from UC Irvine, where he is a professor of informatics and studies, in particular, the potential of video game–based technologies to reinvent education and the way that we learn. Welcome our panel!

APPLAUSE

**D. KRAKAUER** Fantastic! Is this mic working? Can you hear me? Wonderful. I'm just delighted to be here with all of you and with this incredible group of people, for whom I have huge admiration. We're going to be discussing something that's very precious and very rare in the world, and that's intelligence—as opposed to stupidity, which as Harlan Ellison has said is the second most common element in the Universe after hydrogen. So hopefully we'll touch on both over the course of this discussion.

I'm going to kick it right off, and we'll be covering a whole range of issues from intelligence at the individual level, how intelligence is made possible by culture, how we fail and succeed in education, how we represent intelligence to society, the nature of collective intelligence versus individual intelligence, and on and on and on. And, of course, AI!

So there are a few questions that are going be targeted, and then there are questions that are open. I want to start with Vanessa. Vanessa works on cumulative evolutionary processes, and so Vanessa, what makes intelligence—intelligence in our species or any species—possible in the first place?

**VANESSA FERDINAND** Well, first, I want to say that there are many different types of intelligence. For instance, if I gave you six thousand pine nuts and told you to go bury them in the Santa Fe forest over a hundred acres, and then I told you to come back in one year and go get them all, you would not be able to do that. But a piñon jay would entirely "beast" that task because they have incredibly good spatial intelligence. So, if you pick somebody and say, "Tell me what intelligence is," I think in large part what they'll do is describe themselves.

LAUGHTER

**V. FERDINAND** If you ask a species to come up with a theory of intelligence, you're going to get a theory that tells you what their intelligence is. I think we have this human-specific, human-centric view of what intelligence is. And I think that, throughout this panel, in the back of our minds, we need to think about the other types of intelligence that are out there.

**D. KRAKAUER** Yeah, and actually, let me just move straight to Kurt for this, because Kurt's an expert on technology, computation, and gaming in education. Education is nominally that machine that we built to allow for the cumulative transmission of the best of our culture. Why is it so difficult? And, are there prospects out there for

-197-

doing it really well? And if we're going to send education out into space, what form of education are we going be transmitting?

**KURT SQUIRE** Yeah, well, education is built for that, but it's also not, right? So it's built for sorting, which is a lot of what we do, we sort who's going to get which jobs and who isn't. It's also for credentialing, because we worry a lot about that. And then also, I think, it's built for enculturation. I think right now we're seeing an understanding of the role that things like education, or even things like traditional media channels—things I'd love to slag on—have done in enculturating us toward a common culture, so we have a shared understanding of what's factual and what's not. And education, for all of its troubles, is something that's done that enculturation to a degree.

But what about the things it hasn't done? It hasn't really been any good at helping people—well, not "any good," but it's not a system you would pick for developing passion, developing distributed expertise. For most learning systems you'd think, Well, I should probably become better than the teacher, or, I should decide what I want to learn today in graduate school, right? Right now, that's kind of the way it works.

So a lot of my interest has been focused on how we can create new distributed kinds of systems. Or how we can create systems that look a lot more like the systems we have in the real world, where if I want to become an expert harmonica player, if I want to get better, I can do a lot in the real world between videos and affinity groups and so on. And so I think about how we can create our next-generation systems to do that, to then take advantage of some of these systems, but then also maybe do better than them, because we also know that the "Wild West" of these learning systems can be kind of weird. For instance, there are a lot of issues in terms of toxicity and how people treat each other that we may not want to emulate. But, at the same time, there is a whole bunch of opportunity right now.

**D. KRAKAUER** But would you argue that it's just a matter of time before the most intelligent way of becoming intelligent, intelligent in a variety of different ways, is by machine instruction as opposed to human instruction?

**K. SQUIRE** I think it's going to be a hybrid system. I think you're going to continue to see machines play a greater and greater role as various forms of coaches. But we also know that there is something particular to human bonds, to the way we learn from and with each other. There are neurological studies looking at how you receive information and feedback from a valued mentor, who is a kind of person that you emulate and care for. A caring mentor is one of the most powerful things you can be. It's ironic that we don't have them in school, right? How many kids say, "Oh yeah, I go to school because there's a caring mentor there"? Yes, sometimes there is a teacher–student relationship. If you think back to your best experience, that's usually what it was—there was a teacher that you cared for.

-199-

**D. KRAKAUER** Right.

**K. SQUIRE** So I think that you're going to have a hybrid with machines being more like coaches. I also think there are some things we could probably do to have virtual agents that will help coach you and give you a version of that experience relatively easily. But I don't think it's ever going to replace those human bonds.

**D. KRAKAUER** So let's talk to the man on the panel who wants to replace the human bonds, from the evil empire itself, Google, Graham Spencer. Graham and I have had arguments over the years about this. Graham is a total fanboy for neural nets and deep learning. He's been proven thus far to be completely correct, and I've been proven to be completely wrong. And so every task that I thought would not be susceptible to the kinds of techniques that have been developed recently have yielded. And so, Graham, could you maybe introduce some of these ideas of deep learning and how very simple learning, like reinforcement schemes, have given rise to unbelievable success stories?

**GRAHAM SPENCER** Sure. Thank you for that introduction.

**D. KRAKAUER** Oh, yeah. *Laughs.*

**G. SPENCER** The basic idea with something like a neural net or with any machine learning system is that you give it a bunch of examples. So you show it a bunch of pictures of cows and a bunch of pictures of sailboats and a bunch of pictures of cats, and you have a carrot and a stick. Every time it makes the correct judgment, you give it a carrot. Every time it makes the wrong judgment, you whack it with the stick. And over time, it adjusts its internal numbers, its internal weights, and very slowly begins to learn the things that you would like it to learn.

And for a long time it was believed that that system was fine for learning really, really simple things but could never possibly learn to actually identify images or to play the game of Go or to translate language. But what we've found is that through a combination of techniques, learning on the part of the researchers, and a combination of massive, massive computational scale, lots and lots of things that we thought were not learnable turn out to be learnable in that kind of system.

So now we have machines that can translate human language. They can identify images in a picture. They can even caption a picture. You can show a neural network a picture, and it will say, "A young man riding a skateboard through the park." Things that even ten years ago people would have thought were impossible are now definitely feasible.

**D. KRAKAUER** Yeah. And just to round this out, Jonah [*to the crowd*] first of all, I think Jonah writes the best scientifically informed science fiction screenplays in the world.

APPLAUSE

**D. KRAKAUER** I think it's really true. But I want to give two examples of where you've taken diametrically opposite approaches in your work, and I'd like you to comment on these. So your robots in *Interstellar* are not in any sense anthropomorphic but absolutely full of personality and humor. That's one group. And the other set, in *Westworld*, are increasingly terrifying us as the series progresses.

LAUGHTER

**D. KRAKAUER** And I know, because we were talking about this at dinner once, that you spent about a year, out of your world, reading AI literature, just trying to give an accurate representation in a form that all of us can enjoy and really appreciate. So what are your thoughts in terms of communicating what AI is? Clearly you're split on the matter.

-201-

**J. NOLAN** Yeah, I think I find myself torn, as I imagine everyone up here is on almost a daily basis, between this sort of euphoric techno-optimism and an increasingly callous fear that all of our technologies will inherit all of our problems and will carry them along with us.

I talked about this a little bit yesterday when I introduced the *Forbidden Planet* screening, because *Forbidden Planet* was probably the first on-screen depiction of an anthropomorphic robot coming with us to the stars, and it's a bracingly optimistic one. He's strong and capable and loyal.

And then we went through the kind of *Terminator* phase of anthropomorphized robots that, not coincidentally, coincided with the emergence of robots as a genuinely disruptive force in the labor market. We started having automotive factories with robots on the factory floor and people being displaced from their jobs. Unsurprisingly, it tilts a little more towards that primal reaction to a competitive intelligence.

With *Interstellar*, I was interested in doing something that went back to that original idea that if everything goes the way we want it to, we can design these intelligent agents who will come with us to the stars. There is no reason why they can't embody the very best behaviors and attributes that we ascribe to human behavior. They can be loyal, brave, thoughtful, charismatic. And so with *Interstellar*, you kept waiting for the other shoe to drop. But I said no. You can trust these creatures more than you can trust the human crew members.

**D. KRAKAUER** Yeah.

**J. NOLAN** And then, with *Westworld*, we went a slightly different direction.

LAUGHTER

**D. KRAKAUER** We want to ask about what's going to happen, of course, and Jonah's going to remain silent. But I want to talk about this idea, because I think it's very interesting. One of the issues that I know everyone here is interested in—Kurt raised it, Vanessa raised it—is that there is intelligence in the sense that we're all indoctrinated to think about—intelligence as IQ, which is already a ludicrous test, but let's leave that aside. But we don't really talk about balletic intelligence much. We should. And musical intelligence. Maybe we describe Mozart as a genius in the same way we describe Einstein as a genius, which is interesting, but that's about it.

And then there are these characteristics of empathy, as you say, Jonah, loyalty, sanity. Why aren't we talking about artificial wisdom? And presumably if we're going to go out into space, I don't think those aliens are going to be particularly impressed by AlphaGo. It's like, "Here's the best of humanity, really?" They're just going take a torch to the Go board and say, "That was pretty easy."

LAUGHTER

**D. KRAKAUER** So I'm just curious to know what you all think, and Vanessa, you raised it first: Why don't we have empathy for these diverse forms of intelligence? I mean, why are we so obsessed with this one narrow, analytical band?

**V. FERDINAND** Oh, that's a tough question. I would just say it's because we're self-centered. That's the only thing. I don't know where else to go with that.

**D. KRAKAUER** What about other animals? What about the intelligence of other animals and what we've learnt about them and how that might enrich the debate?

**V. FERDINAND** I would assume people and scientists would be interested in these other animal intelligences, so far as they're useful. I know there's been a lot of attention paid to collective intelligence in ants and termites because termites seem to be doing something that's cumulative culture in a way. I research cumulative culture. I want to point out that there are two very different senses with which you can understand cumulative culture.

-203-

> I don't think those aliens are going to be particularly impressed by AlphaGo. It's like, "Here's the best of humanity, really?" They're just going take a torch to the Go board and say, "That was pretty easy."

The first one is just the accumulation of things. If you look around us, you see tons and tons and tons of things. It's a very materialistic manifestation of what we think intelligence is, and that's the one thing that we can really point to and say is unique to humans: we've got a lot of things. But the other sense of it is, well, my society may have $X$ number of person-hours that we can spend making stuff, and then there is the sense of improvement upon what somebody else has already made. That's cumulative in that it produces outcomes that no one individual could have created on their own. So I could just choose to use my person-hours on reinventing the wheel, or on adding something to the wheel. Termite mounds are a good example, because no individual termite makes the whole mound. They inherit this modified environment and add to it, and it produces all these emergent properties that come from what a termite mound can do as a collective.

So in a sense I'd say that cumulative culture really underlies collective intelligence. If we want to not talk just about human intelligence but rather think more generally about what is special that humans or other organisms like termites are doing, or what other organisms on other planets are doing, they might be tapping into this kind of collective intelligence. The stuff that I've learned, I can package into some informational form. I can package my knowledge and give it to somebody else before I die so that knowledge doesn't die with me. It stays out there; it's public domain and could then be used to take more steps beyond.

**D. KRAKAUER** But, since you've given the rosy picture of collective intelligence, let's look at the slightly less rosy version, which is horribly malicious chatbots or the—

**K. SQUIRE** Go to Tucson right now, right? The thing going on in Tucson right now is crazy.

**D. KRAKAUER** What is it?

**J. NOLAN** What is that?

**K. SQUIRE** Oh, there's a group of people there who are convinced that there are homeless people trapping children and enslaving them for sexual purposes. And right now they're convinced this is true. There's no evidence; this is not happening.

**D. KRAKAUER** This is just a little bubble . . .

**K. SQUIRE** Yes, there are thousands of people who are dedicating hours to this and yet they can't see the obvious logical problem.

**D. KRAKAUER** But that's the thing, okay, so let's talk about that. And I know Graham has worked a lot on issues of bias in algorithms. So if the data that's being collected is being generated by fallible human beings, with all our prejudices, it's just going to reflect it right back at us in a more efficient manner, right? So what are your thoughts on that? I mean, what is the negative side of Vanessa's comment of collective intelligence in the new tech world?

**G. SPENCER** It's a huge problem. One of the best examples is in the field of machine translation, where a computer will take a whole bunch of snippets of text from the United Nations, for example, where you have the same text in two different languages. And it will look for matching segments of text and it will build up a giant database of matching segments, and then when you feed it a big sentence, it tries to find the best matches and give you a translation.

-205-

But quickly people noticed a big problem with that, for instance, when you take a language like English, which has gendered pronouns, and you translate that into a language with nongendered pronouns, and then you translate it back. So if you say, "He is a doctor; she is a nurse," when that gets translated into the nongendered language, it's "They are a doctor; they are a nurse." And when it comes back to English, it's right back to "He is a doctor; she is a nurse." So it's preserving the gender biases of the source language in a way that people didn't really expect. Of course, it's obvious in hindsight, but figuring out how to address that is extremely difficult. I don't think anyone has a good answer right now because it crops up in so many different ways.

It crops up in facial recognition, too, where the databases for facial recognition are usually 80 or 90 percent white, so they have a 5 percent accuracy on white male faces but a 30 percent accuracy on black female faces. And in part that's because many of the researchers who do that work are themselves white males. And partly it's because the database that they use is dominated by white male faces. Almost everywhere you turn you find these issues and I don't think anyone knows of a good solution yet. But, yes, it really does crystallize and reflect back the biases that we already have.

**K. SQUIRE** How are you thinking about the ethics here, once these things exist? As in, how could they be used for things you might plan for and things you might not plan for? Because that's been one of my concerns about all of the things that we create: Once they're out there, who knows what's going to be done with them?

**G. SPENCER** It's a huge problem. I mean, there was a very controversial paper written, I think, just within the past year, where someone took a bunch of faces of criminals and noncriminals and then tried to build a machine-learning algorithm using a vision recognition system to identify whether someone's a criminal from their face. And of course —

**J. NOLAN** That's just phrenology.

**G. SPENCER** Yes, exactly.

**J. NOLAN** Phrenology 2.0, right?

**G. SPENCER** Exactly, and anyone, anyone can look at that experiment and realize how incredibly biased and wrong that is, but anyone can go download the tool kit to do that; it's out there. People have done this to detect sexual orientation also. Again, this is opening up an incredible set of problems, and the predictions are almost certainly completely inaccurate, but you can build the model and you can run the model and then you can believe the model, and that might make you do all kinds of things. So it's really happening now.

**D. KRAKAUER** Can I ask . . . Kurt and Jonah are, in different ways, reaching out. In a way you're both educators. It's an odd thing to say, but I think it's true. You are exploring the hybrid domain, right? That is, how a form of intelligence we think we know—our own—is interacting with one that we don't—one that we created—and so I'm just curious to know what your thoughts are on those hybrids and how that plays out in different ways in your work? I mean, for Kurt, does the machine have to be human for us to empathize with it? Or could it be unbelievably different? In fact, that might be preferable, actually.

**K. SQUIRE** Yeah.

**D. KRAKAUER** And, in your experience, Jonah, writing about these things, do they have to be human values? I mean, are there values we could admire and appreciate that are vastly different? I want to just riff on these kinds of ideas.

**K. SQUIRE** Well, I think anyone who's played, say, *Final Fantasy VII*, all of the gamers out there know that you can care a lot about synthetic characters very quickly and easily, and we do in games all the time. It's super simple. You have a game, you let your character do something to help you, and then you usually feel a bond pretty quickly. We've been exploring ways to do this for other purposes.

Right now we're working on creating virtual agents to help kids -207- monitor their attention and their executive functioning. So we're creating ways to help them become better able to manage things like their attention. But I'm thinking, Oh gosh, we're inventing something that could be scary. I do worry about who gets those data. What do they do with them? Do teachers get it? Do parents get it? What happens when devices know your emotional state?

One thing I'm committed to, though, is giving people the keys to that, in the same way that I think we should more broadly be able to control what happens to our Facebook and Google data, which is something that we as consumers have kind of given up. But it happens even in other systems, too. As a student, you could imagine saying, "I should be in charge of what happens with my test scores and what the meanings are," and "I should have the ability to be more of an advocate for myself in our system." So that's something we've been trying to do. We've been trying to create social systems that get around all of that so that we encourage kids to be the ones to make the case for what they think the meaning of a score is, or what they think the meaning of this or that is, so that they're arguing with their data toward something. But I'd be lying if I said that we have it figured out.

**J. NOLAN** I want to jump in on this question of empathy, which feels pretty foundational. We're going to need to build it into any form of AGI [artificial general intelligence] we contemplate. And for us, not on a theoretical level but on a market-driven level, we had to grapple with the problem of empathy. We're doing a show about robots, and your show lasts only as long as the audience cares about the characters. We struggled with it on a technical level in the first episode. During the notes process, people would say things that you couldn't believe they would say. The whole project became a kind of acid test for the people we were working with in frightening ways. One producer at one point said, "Well, yeah, they get beaten up and assaulted and attacked, but if they don't remember any of it, does it really matter? Is there really a moral component to that?" And my wife, who worked in the DA's office and has dealt with family violence of children who wouldn't remember the things that were done to them, had to point out that it was still a crime.

Largely, we have an incredibly talented cast, and the actors who are playing hosts, who are the robots on our show, they would do an awful lot of acting. But we would go in and do a bit of subtle manipulation to their performances. We would slow down the way their facial muscles recovered from a blink. We would make the eyes move slightly independently. It was really subtle stuff, and we had to be very careful with how much of it we applied. Humans are really weird when it comes to the question of empathy. I can draw a smile on a volleyball and build a very successful Tom Hanks movie out of that where you feel really bad when the volleyball gets swept away by the ocean.

**D. KRAKAUER** I didn't feel bad at all.

**J. NOLAN** Yeah, I didn't care. I didn't care. If it was a basketball, I might have been more interested. But no, so humans are weird. I can trick you into applying empathy to something that does not deserve it in the slightest. So I have a daughter and she has lots of stuffed animals, and she would feel terrible if one of them came to any harm. But, we're very bad at applying it, even to our fellow men, right? It's easy for us to make a distinction.

So, getting into the mechanics beyond artificial intelligence, how do we sort things out? How do we understand things? This idea of artificial wisdom, artificial sanity, these larger components of what it is to be intelligent, which includes empathy, they're something I barely understand in the human creature.

**K. SQUIRE** If we design that artificial wisdom, maybe that's the thing that'll kill the humans finally.

LAUGHTER

**K. SQUIRE** It's like, that's it. You're out.

**D. KRAKAUER** So let's talk about the elephant in the room. We're here at the InterPlanetary Festival, and it's ideas and it's art and it's positive and constructive, I hope. And yet we're living in a world which, at the present time, seems to be so riddled with prejudice and stupidity and intolerance. It always has been, of course, but we seem to be going through a bit of a bad phase. But, you know, we're making so much progress in so many ways, right? Every day you pick up the newspaper, and something extraordinary has been discovered. Someone has done some really amazing work. And so I wonder, what are we not doing? Where are we failing and why? Why isn't the culture, to Vanessa's point, why aren't we doing this cumulative culture thing? Why aren't these ideas getting out there? Why isn't there an honest debate to reckon with facts and come to a sort of better resolution? What's going on in the world? Why is it so stupid?

-209-

LAUGHTER

**V. FERDINAND** Maybe I'll half answer that by hearkening back to what Graham was talking about, biases. And I'll give you some theoretical background for that, as a cognitive scientist. Culture evolves by adapting to people's minds. Culture is something that replicates by passing through cognitive systems, through one mind, through another mind, through another mind, like that. And, in that sense, cognitive biases are selection pressures on culture. So if you can't learn something for whatever reason, there's no way that you're going to be able to teach it, and it's going to fall out of your cultural domain.

So you really can't expect culture to solve these problems for you. You need to change the landscape that culture is adapting to, which is our mind. And that's why language engineering, for example, is so bad. It doesn't work.

There's a really good example in the word *Ms.*, which was supposed to be a gendered way to refer to a woman without disclosing her marital status. I saw some application forms from the '70s that had Ms. on there, and this was a new thing. A lot of people didn't know what it was, and they were like, why would someone not want to disclose their marital status? And I actually saw some applications that said, "Ms. (divorced)."

LAUGHTER

## Stupidity . . . where do you even start?

**V. FERDINAND** Yeah. And nowadays, of course, we know, and we don't use it like that.

There's a lot in gender-neutral language. *Chairperson* is another really good example. *Chairperson* should not specify the gender of a person who's chairing a meeting. But if you go on Google Ngram, and you should—it's a big database of word frequencies over time. If you go to Google Ngram and you type in, "He is the chairperson," you'll have a flat line for "he is the chairperson." Millions, billions and trillions of words in English, and nobody's ever said that phrase. But if you type in "she is the chairperson," that phrase has been rising along with *chairperson*. This is a gender-neutral word that people invented, but then they released it into a population of biased minds and it just finds its niche. So we really need to change our minds; we can't expect culture to do anything for us, in this sense.

**D. KRAKAUER** Right. Right. That's great. Stupidity. *Laughs.*

**J. NOLAN** Stupidity . . . where do you even start? We've talked a little bit about this before. I'm fascinated by the idea of the Flynn effect, which suggests, for those of you not familiar with it, that over the hundred or so years in which there's been large-scale intelligence testing, they have to renormalize every ten years because the population seems to be getting smarter.

Now, there's an enormous amount of questions underlying that. Are they just getting better at doing the test? What do we mean when we say they're getting smarter? What do IQ tests even measure? But let's assume that on some level the Flynn effect means that we're getting smarter. It's hard to imagine. You see a number of different explanations for what nutrition is, right? The green revolution, the idea that there's far more access to nutritive food. It's possible. I do tend to think and wonder—again, knowing that humans always tend to see it through their own lens—whether it's technical or something connected to our technological expressions of culture. It seems to neatly connect to the rise of television. So now the idiot box, ironically, might be part of this broadening of the ability for cultures to exchange information, for people to ideate back and forth through technology. If there is any connection to technological sort of underpinning to the Flynn effect, it's—

**K. SQUIRE** It's video games! Sorry.

**J. NOLAN** It's clearly video games. You asked a second question, David. I was very interested in a talk you gave in LA last year, and afterwards we talked about the idea that the singularity actually began seventy thousand years ago with the advent of spoken language and written language and the idea that intelligence is a shared phenomenon, that we ideate back and forth.

And, for me, the most urgent question right now is—holding open the possibility that technology impacts that general level of intelligence—what's happening with social networking? If that's where most of us are getting our information, if that's where most of us are now exchanging our ideas, well, those structures have been brought into place by market forces, right? So there's this larger question of what's ethical? What's sensible? What contributes to the overall good of it? I think that has become a very urgent question.

You can look at social networking as a dry run for artificial intelligence because largely it's the same two companies who are driving both forward. Facebook is the second biggest investor in artificial intelligence behind Google. If you look at social networking as a dry run, I think it's been a disaster. I mean, just look at the last year. Every day this week there's another article about Facebook. So to that larger question about the technologies we make, are they designed the right way to further the overall intelligence of the human species?

**D. KRAKAUER** That's really interesting, and it gets to this notion of collective intelligence—something that many of us are interested in at SFI, and it's not obvious, right, that to create a collective intelligence you just create lots of intelligent components. I mean, ants arguably have much more impressive collective intelligence than we do in some sense, but they're not individually more intelligent. So the mappings are rather complex.

Now I know that we're running out of time, so I'm going ask you this one final question. It's a one-word answer. And that is, if we're going to send into space a species, animal or plant or microbial, that best represents the intelligence of evolution on the planet Earth. This is a tough one—

**J. NOLAN** Ants.

**D. KRAKAUER** Yeah, but you're going to have to send a colony, Jonah. Which species or individual representative of a species would you send? And I'm going to start, because I know you have to think about this. It's obvious: it's a cephalopod, obviously, probably a giant Pacific octopus. That's mine.

**V. FERDINAND** I'd say a spore, because it has a lot of staying power.

**D. KRAKAUER** A spore?

**V. FERDINAND** Yeah!

**D. KRAKAUER** Of what species?

**V. FERDINAND** It's robust to everything!

**D. KRAKAUER** What kind of spore?

**V. FERDINAND** Or I'd send a bacterium or something that's proven that it's good on Earth, you know?

**D. KRAKAUER** All right. Okay. Done. And take them out when it hits them. That's right! Graham?

**G. SPENCER** I'm going to be totally speciesist and say a human. I'm going to do it.

**D. KRAKAUER** All right.

**J. NOLAN** I'm going to go with Douglas Adams on this one and say dolphin.

**K. SQUIRE** *Sighs.*

**D. KRAKAUER** Nice!

**J. NOLAN** You can't blow up a planet with dolphins.

**K. SQUIRE** That was mine. I could pick a different species maybe. The bottlenose? No? Okay, the ant, I guess, for the collective ability to do work. Or bees! Either one. I said ants, so I'll go with ants.

**D. KRAKAUER** That's great. Well, thank you very much to the panel. Thank you all very much.

# INTRODUCTION:
# SOCIAL & ECONOMIC ENGINEERING

*Well, societies without a plan, that was history so far; but
history so far had been a nightmare, a huge compendium
of examples to be avoided. . . .*

*"And with our work," John continued, "we are carving out
a new social order and the next step in the human story."*
—KIM STANLEY ROBINSON, *RED MARS*

We humans are at an intersection. The convergence of ideas and
events at this intersection is just coming into focus but is significant enough that the intersection might be, for human and possibly planetary history, a critical point—a "point" at which a small
perturbation can cause the system to shift to a new state. A state
that is, for example, either still organized, like ice—but in a different
way than the current state—or disorganized, like a gas. This intersection might, more significantly even, be an origin point. Before
we can stretch the limits of our collective imagination to consider
what future history could originate from this moment in time, it is
worthwhile to consider the nature of the convergence.

Machine learning and artificial intelligence have been around for
a long time (that is, longer than the media hype would suggest)—
since at least the 1950s, when, at the Dartmouth 1956 AI conference,
the field is said to have been founded. The pace of research is now
increasing exponentially, if the number of machine learning and AI
papers uploaded to the physics arXiv is any indication.

The 1990s mark the start of the controversially named "big data" era
on two fronts: the biological front, with high-throughput genomics
data, and the "digital" front with the rise of personal computing

(and, hence, the opportunity to track, record, and quantify individual behavior on a large scale) and individual interactions *vis-à-vis* social media. With these developments, we have available unprecedented microscopic data on interactions, behavior, genetics, and physiology, that can be harnessed to study the mapping between individual-level behavior and group or collective behavior, whether of cells, brains, or societies.

Coincident with these developments, the science of micro to macro in adaptive systems is blossoming due largely, I would say, to the rise of complexity science. The contributions of complexity science have been philosophical, conceptual, and technical. Philosophically, complexity science, borrowing from physics, presupposes that with the right lens it is possible to discover organizational principles that are scale- and substrate-independent. Conceptual contributions include moving away from substrate-specific questions to an emphasis on problems that reoccur across adaptive systems—for example, those relating to information processing and computation, robustness, communication and coordination, emergence, scaling, learning, and evolutionary dynamics. Technical contributions include techniques for studying micro to macro mappings by combining insights and approaches from statistical mechanics, theoretical computer science, and network theory; information theory for recasting evolutionary dynamics in terms of changes to mutual information; maximum entropy approaches and information theory for quantifying how collective or decomposable a system is (which is critical to a theory of control or intervention); and dimension-reduction and coarse-graining techniques to identify the dominant causal contributions to macroscopic change and hence build a theory for regularities observed at the macroscopic scale. One of the best examples we have so far of a successful identification and derivation of lawlike behavior in adaptive systems is the work on metabolic scaling of Geoffrey West and colleagues, which is accessibly discussed in West's book, *Scale*. As Freeman Dyson put it in a provocative but in some ways maddeningly incorrect review of Geoffrey's work and, more generally, SFI's research program, we (humans) are somewhere between Galileo and Newton in our understanding of adaptive systems.

Two other developments at this intersection are climate change, with its rapidly accelerating pace, and a re-injection of energy into space exploration—in particular, getting to Mars, largely through research and development by private companies like SpaceX, Blue Origin, and Virgin Galactic.

These five factors—AI, big data, a blossoming understanding of micro to macro in adaptive systems, climate change, and space travel—are linked, but not in a trivial way. A longer essay and a lot of thought would be required to work out the relationships. For now, the convergence is an observation with at least some clear if not yet concrete implications. With respect to this panel, the most relevant of these implications is that for the first time in human history a quantitative science of social, environmental, and economic engineering looks possible.

Humans have been attempting to engineer social outcomes since the dawn of cultural history. As I mentioned in the panel, there are many great examples. In his book *Priests and Programmers*, anthropologist Steve Lansing describes how a Balinese water temple system emerged in the ninth century to optimize planting cycles and water distribution. A Rube Goldberg–like voting process for electing the Doge, described beautifully by John Julius Norwich in *A History of Venice*, was invented somewhat cooperatively by rival Venetian families in the 1500s to help prevent the process from being gamed.

These examples, however surprisingly elegant, differ from canonical examples of engineering and manufacturing plants and cars and spaceships in that there are no blueprints for social systems, no rigorously quantitative way yet to identify targets of interventions that will reliably produce or control change. Rather, the history of human social, financial, and ecosystem engineering is based largely on intuition. As Isaac Asimov wrote in an essay in *The Planet that Wasn't,*

> "People are entirely too disbelieving of coincidence. They are far too ready to dismiss it and to build arcane structures of extremely rickety substance in order to avoid it. I, on the other hand, see coincidence everywhere as an inevitable consequence of the laws of probability, according to which having no unusual coincidence is far more unusual than any coincidence could possibly be."

-217-

Perhaps consequently, the majority of attempts to engineer adaptive systems have been disastrous or impotent, especially those that did not have the benefit of developing organically over a long time period, as self-organization and long timescales can sometimes compensate for cruddy intuition. We might call the past history of social engineering reactive. The future can in principle be proactive.

..............................................................

## Humans have been attempting to engineer social outcomes since the dawn of cultural history.

..............................................................

The behavioral maps we will be able to build with the vast microscopic data now being collected, developments in AI and complexity science, and motivation from the desire to get to Mars and control the climate might allow us to find and quantify the hidden regularities in our social interactions—regularities that, thus far, we have been unable to measure, or which may have been invisible given our myopic perception and intuition-dominated reasoning. Not to mention the fact that most of the adaptive systems we want to influence are complex, with multiple time and space scale, heterogeneous actors, and learning, as well as evolutionary dynamics.

If we can successfully infer the rules and strategies individuals use to guide decision-making, we will have a robust starting point for building predictive simulations of social outcomes at the societal level. Such simulations will also allow us to test alternative futures and give a quantitative, empirical basis for our intervention decisions.

The potential power of this approach seems obvious. One only has to look at the huge investment into data collection by corporations like Google and Facebook and all of the third-party companies—data merchants—that deal solely in data sales, or to China's social credit program. But, amazingly, just ten years ago, there was little discussion outside of science-fiction novels of this growing reality. The public as well as many scientists scoffed at it as a pipe dream. The 2016 Facebook election debacle—even if it is overhyped—is an example of just how poorly we collectively anticipated the change from reactive to the beginnings of a proactive, quantitative social engineering, and how rapid its initial stages might be.

Will there be a giant leap forward in proactive social engineering allowing the orchestration of precisely engineered individual- or societal-level outcomes? Probably not. Two reasons why this is unlikely are the stochastic (random), rather than deterministic, nature of human behavior and the stochasticity in the process by which behavior combines to produce society. Even if scientists had the best data and methods at their disposal, complete prediction would never be possible because of the character of adaptive systems. Adaptive systems are error-prone computers making estimates based on finite, imperfect data, and they are subject to changing environments. One might respond to this pessimism by suggesting that social engineering could reduce behavioral variance and hence eliminate much noise. But that view is naïve, and brings us to a second reason why social engineering will never rival standard engineering in its predictive power.

Adaptive systems are just that—adaptive; their actors respond strategically as the system and environment changes, in evolutionary or learning time, meaning that metrics used to gauge adaptation, once they become targets, can cease to be useful metrics. This is Goodhart's Law, and it applies when the timescale separation between the microscopic and macroscopic is too small. Social engineering may in fact make social systems more, not less, complicated and predictable if these engineering arms races get out of control. Perhaps a way around this is for social engineering to focus on process over orchestrating specific outcomes like the degree of inequality.

What is much more likely than tight control over the future is coarse but robust prediction based on an understanding of dominant causes at the mesoscale. This is illustrated by a recent study in *Science* by Nicolas Bain and Denis Bartolo of the collective motion of marathoners. This study found the large-scale motion of the runners could be predicted without knowledge of individual interaction rules if the crowd was modeled as a fluid. The question for social engineering is: can we build societies that are like fluids, so that we can predict and control aggregate behavior without having to know, or care about, what the individuals are doing? Assuming that such models work even when the range of individual behavior is large and varied (unlike in marathons), this would resolve many problems concerning individual–collective trade-offs, such as privacy and autonomy, and might allow a societal engineering to which we can all contribute cooperatively and adversarially, as our local needs dictate.

-219-

A little imagination has gone into this introduction. If we are to build an interplanetary civilization and want to make use of our growing capacity for social engineering to do it, we are going to need a lot of thinking outside the box. The purpose of the InterPlanetary Festival and this panel in particular is to accelerate that discussion.

*—Jessica Flack*
*Professor, C4 Director, & Chair of Public Events*
*Santa Fe Institute*

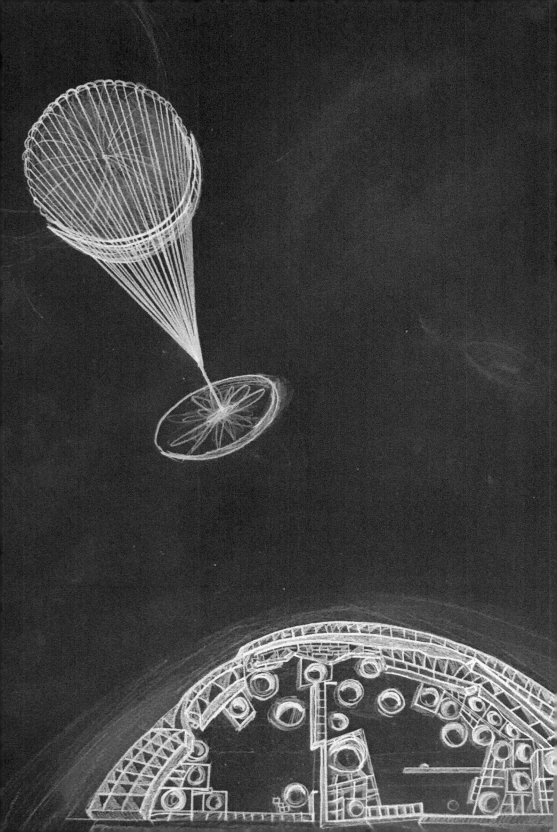

# SOCIAL & ECONOMIC ENGINEERING

*David Krakauer introduces the panel, moderated by Jessica Flack
and featuring Cory Doctorow, Robert Gehorsam, and D.A. Wallach.*

**DAVID KRAKAUER** I'm delighted to introduce this next panel
on social and economic engineering of the future, hopefully with a
very positive connotation of engineering. Let me just briefly introduce the panelists. Here on my left is Jessica Flack. She is a professor
at SFI who works on collective computation—so how it is when
many individuals come together they produce emergent properties,
whereby you could treat society almost as if it were a circuit board,
respecting, of course, the agency of individual components.

Sitting next to her is Robert Gehorsam, who started his career as a
book editor at Simon & Schuster, and then saw the light and moved
into virtual and augmented reality. He's worked with a huge number
of very well-known companies: Prodigy Services, Scholastic, Sony,
Viacom, and, most interestingly, he was the director of the Institute
of Play. And Rob has a very intriguing attitude towards the sort of
iconoclastic treatment of discovery.

Sitting by him is the great writer Cory Doctorow, yay!

APPLAUSE

**D. KRAKAUER** Cory is many things: he's an activist, he's someone
who has thought deeply about the modern world in which we live,
the open source movement, the control and dissemination of information, and his novels and books and blog posts, and so on reflect
those interests. His most recent and wonderful novel is *Walkaway*,
which means, yes, you can, I guess.

**CORY DOCTOROW** Well...

**D. KRAKAUER** But also *In Real Life* and *Information Doesn't Want to Be Free*, which is a twist on a famous false remark.

And sitting next to him is the wonderful, the inimitable, the only, D.A. Wallach, whom you've met, of course, introducing some of these InterPlanetary sessions. D.A. is hard to describe. He started off learning several languages; he spent a lot of time in Africa, working in a number of different communities, and then he became a rock star. He's a talented musician. He appears in the film *La La Land*. But then, something happened—now he's a venture capitalist. I don't know, man, you might have to explain yourself! He's doing venture capital for good, and he's become an extraordinary expert in molecular biology, nanotechnology, and the interface between those principles and inference and machine learning.

So this is an amazing panel—take it away, Jess.

APPLAUSE

**JESSICA FLACK** Okay! Welcome, everybody. I hope you're all having a good time at InterPlanetary. What I want to do is start with a few minutes, maybe two or three minutes, of context to ground the discussion. And the first thing I want to say is that human history is replete with examples of attempts to orchestrate social outcomes and to engineer social interactions. A great example comes from here in New Mexico itself: Chaco Canyon. Another example is the Balinese water temple system that emerged in the ninth century in Southeast Asia. There you have amazing, iconic, stylized monuments and the evolution of all sorts of elaborate rituals to control planting cycles and water distribution.

Another example, and of course this is a very different kind of example, is voting systems. A really interesting one comes from Venice; it was an incredibly opaque, Rube Goldberg–esque voting protocol that was invented by feuding Venetian families in the 1500s to make the election of the Doge tamper-free. So these are all examples of what I would call social engineering. Now, of course, if you are an actual engineer, you might balk at those examples, because social engineering is not particularly precise, or it certainly hasn't been, and it's not at all like designing a cell phone tower or a space ship, or a water mill for milling grain. It's very different: there are no meticulously drawn blueprints; there are no objective tests we can use to decide whether a particular social intervention is actually going to work to change things. In fact, it's largely only based on human intuition about cause and effect, and then trying to manipulate or change things based on our understanding of cause and effect. For example, if the crime rate goes up, we build more prisons and increase our police force. It's kind of post hoc, retrospective, and based on intuition.

-223-

And so you can say that so far in human history, social engineering has been what we call, or what I've called in our work, largely *re*active instead of *pro*active, right? And the reason for this is that we haven't really had realistic, fine-grained data on human behavior. But, as you all know, as you have gained a glimmer of, no doubt, with the Cambridge Analytica–Facebook debacles, this is beginning to change. It's not clear how it will change yet, but it is definitely beginning to change. And we are getting, for the first time in human history, incredibly detailed data on human social interactions; we're getting great data on humans in groups; and we're getting data on how collective decisions are made.

So the world is about to change, I would say, and the question is, with this digital technology and that great data, are we in a position to engineer social outcomes from the bottom up with quantitative principles? This is an important concept, one that we can certainly employ on Earth now, and one that we can maybe take off Earth as we explore new worlds. Given that we have this opportunity, maybe the time has come to really think about it, to think outside the box, to think about what would be possible if we weren't limited by our own imaginations.

So that's what we're going to talk a bit about today. Before we jump into the discussion, these guys are going to say a little bit about their own backgrounds, and maybe make a provocative statement or give their perspective on this issue. Let's just go down the line.

**ROBERT GEHORSAM** Okay, can you all hear me? So, as David said, I was working in the book industry, and then I "saw the light" and switched into digital industry, specifically games, and for many years I was interested in games as an emerging art form, simply as a form of fiction. But, for the past fifteen years, it's become clear to me that they are a form of social experimentation, especially as games became social and connected. And so they've become, for me, an interesting way to do social engineering, and to gain social understanding, at different timescales. So while we have the real-time world of Cambridge Analytica, it's possible to do experimentation, to think of game-like environments as terraria for exploring the future. That's of great interest to me.

**C. DOCTOROW** I'm a science fiction novelist, and that means that I often think about the predictive power of different kinds of imaginative exercises. And I've come to the conclusion that science fiction has no predictive power. I think that science fiction writers are basically Texas marksmen: we like to fire a shotgun into the side of a barn, and then draw a target around the place where the pellets went in and tell everyone what a great shot we are. But science fiction does some other interesting and useful things. It's kind of obvious on its face that science fiction is a really good inspiration or warning. You know, all appearances to the contrary, George Orwell did not intend *1984* to be a manual for statecraft; it was meant to be something of a warning, but people got inspired. Paul Krugman became an economist because he thought Hari Seldon in Asimov's books seemed like a cool dude.

But that warning and inspiration is itself a first-order effect; the more interesting second-order effect is the diagnostic power of science fiction. So if you want to know what a science fiction writer is secretly worried about or hopeful for, read what they've written. If you want to know what the world is secretly hopeful for or fearing, read what they're reading—find out what's on the bestseller list. Mary Shelley's book is two hundred years old, and we're still reading it probably for very different reasons than when they were reading *Frankenstein* two hundred years ago but, nevertheless, the fact that we're now really worried about runaway technology becoming our master instead of our servant tells you a lot about where we are today.

**D.A. WALLACH** Since David framed my transition to venture capital investing as a little paradoxical, I guess I'll just defend it for a second, and maybe that's a good way to explain myself. Broadly speaking, if you're investing other people's money, the goal is to give them back more money than they gave you. And I tend to think that there are two ways of doing this. One is that you identify things in the world that you think are extremely unlikely to change, and you place bets on those things. That's probably the best way to reliably make money. And then the second is to make predictions about what things, if they existed, people would spend a lot of their money on. And that is, I think, inherently a project of social engineering.

-225-

So the companies that I try to back are companies that have a new idea about how we should do something, and they're either vindicated or discredited when they actually give people the option to say yes or no to that experience. Over time, of course, my hope is to get better and better at predicting this, but the more I've done it, the more obvious it becomes that it's a really difficult thing to do. Predicting in general is really hard, and, as Jessica said, social engineering is extremely imprecise.

However, I guess what gives me hope is that all of these changes in human behavior tend to begin with some act of creativity—whether that's the proposal of a new scientific theory, or an explanation of something, or by creating an experience that people can participate in, and then choose to. It just doesn't happen unless people give it a shot. So there needs to be a source of new ideas; and for those new ideas to become important in our society, driven as it is so much by economic activity, those have to be ideas that people can participate in economically.

**J. FLACK** Awesome, guys! So let's actually dive into that a little bit, and think about this question of limits. I think one question we might ask is, are the current social institutions that we have the best there is, for example, the ones that we've seen over human history? Or are they a consequence of the sort of constraints that Earth has actually placed on us, natural ecology and so forth? Or our brains maybe? And to what extent are they also the limits of our imagination?

**R. GEHORSAM** I'll go first. I think it's self-evident that they're probably not the best that there ever will be. They're defined by conditions here, past and present, and when we talk about social engineering as a distributed interplanetary society, there are going to be new conditions. The question is, will we reflexively seek to impose old models on new conditions in a context where the cost of failure is very high, where the settlements in fragile environments will collapse if it's wrong?

**J. FLACK** So there is this trade-off between doing what you know works and being exploratory. And, of course, you sort of know that what works might work for our brains, for example, in the way that we interact with each other, but maybe not in a new ecology.

**R. GEHORSAM** Right.

**J. FLACK** So how do you balance that "getting it wrong"? Especially when the cost, like you say, if we go to Mars or something, is really high?

**C. DOCTOROW** Well, you know, I think that when we think about engineering problems, we can cleave them into two halves. One is whether we've designed a system that works well, and one is whether we've designed a system that fails gracefully. And when you say something is "best" or "worst," or "is it as good as it can be?" and "what are its constraints?"—well, for a system that works very well but fails to adapt to changing circumstances, its lack of graceful failure is probably a deal breaker. It doesn't matter how much general prosperity you produce during the golden years if you are all drinking your urine and digging through rubble, looking for canned goods, in ten years. You want a soft landing, as well as a great voyage, on the way.

My concern about any or all of the ways that we contemplate off-planet civilizations or off-planet societies is that they all seem to have these authoritarian underpinnings. Either they're conceived of in an explicitly military construct—for example, we have words like *captain*, as part of the construct. And you see this in Heinlein and other science fiction of the age, where oftentimes there's someone who is literally wearing a sidearm and shouting, "It's lifeboat rules!" whenever someone questions their authority. Because, you know, "we're all in a lifeboat, we don't know how much longer we've got. I'm the only one who knows how to steer the lifeboat. You do what you're told, or I shoot you and toss you overboard." And whether or not there may be times in a lifeboat when you need to do that, for a certain kind of person, sinking the ship so that you all get in the lifeboat and you get to shout, "Lifeboat rules!" whenever anyone disagrees with you might be an awfully attractive proposition.

Or we could be talking about Iron Man sending us all to Mars in our Teslas. That is a guy who by all accounts is the world's shittiest boss.

LAUGHTER

**C. DOCTOROW** And the idea that we'll just all subject ourselves to his management style, it's a kind of off-world, no-escape, Sartrean vision, or the original vision for Disney World, where you'd have a dome and everyone would live there, and Walt Disney would replace the Bill of Rights with a set of employment terms that would involve not swearing or not having a mustache. Everyone has to drink Tom Collinses instead of caipirinhas or something else, those kinds of aesthetic ideas that rich weirdos come up with and declare to be natural laws.

APPLAUSE

**J. FLACK** So part of this, I guess, is having a fluid organizational structure. I think we had some discussions with NASA once, where they were actually talking about the problem of getting to Mars over ten months. Part of it is a standard NASA mission, where you need that command style and organization that they drew from the military, but part of it is a science expedition. And they need to be able to shift, over those ten months, between these two very different organizational structures. So how do you design that? What kind of team is the right team that can go between these two very different states?

**D.A. WALLACH** I think a really interesting real-world example of social engineering is the American civil rights movement, which over decades navigated between a number of different organizational structures. So you had grass-roots, decades-long legal campaigns going on to try and elucidate the ways in which the law was being misinterpreted to produce social outcomes that were obviously unjust to the people who were critiquing them. You also had galvanizing leadership in the form of folks like Martin Luther King, Jr., or Malcolm X, or others who appeared to be almost dictatorial figures, but who stood on the shoulders of thousands of grass-roots volunteers who were actually changing people's perceptions.

And the other thing that I think is really interesting about the civil rights movement, or the suffragette movement, we could talk about movements elsewhere in the world, is that there were efforts at social engineering that didn't begin from some idealist vision of the future. They began from a place of identifying obvious problems and injustice. And when I look out at the world, I think there are obviously so many mistakes in the way we do things, that given our limited time, better to start with those. Let's triage the things that we know don't work, and just make them a little bit better at least. I think where we often erred in the past, as we have thought about ourselves as doing social engineering, is by saying, "Throw out everything." And we have some picture in some guy's book of exactly what it should look like.

**J. FLACK** That's a nice example, D.A. Tell me a little bit more about it, because what it sounds like is that, of course, there wasn't a big vision for the civil rights movement that was conceived at the beginning. These personalities and aspects sort of filled a niche that developed as the civil rights movement grew. So it's reactive, in the sense that I was suggesting in the introduction, like many of the examples that we have, instead of proactive in others. There's an idea in evolution called *niche construction* where basically organisms, humans included, go out and modify the environment. They change the selection pressures they're subject to, and therefore they sort of control their future and can predict the future a little bit more. That's very proactive. Are there any elements that we can take from the civil rights example that have that kind of proactive feature?

-229-

**D.A. WALLACH** I think you get into causality in some sense because the question is, if you twist this knob in society, what happens? There are certain laws that are obviously pretty clearly connected to their outcomes. So if, for example, you have a law that prohibits black people from voting, you can be relatively confident that more black people will vote if you make it legal for them to do so.

And so I think you start with that kind of low-hanging fruit, and then it gets more complicated when you get into issues that currently surround issues of racial justice in America. For example, what different sentencing laws will do in the way of outcomes in a specific neighborhood, where the product may be the result of a lot of different processes that that's just one part of.

**C. DOCTOROW** I think, maybe, the distinction that you guys are getting towards is tactics and strategy. I think that there was a vision, right? MLK was a socialist who believed in a kind of egalitarian outcome; that was the thing that he was headed towards—the thing that the long arc of history was bending towards, in his view. And then he has a bunch of tactical ideas for how to accomplish it, as do his colleagues. And I think that this is where our discussions about, "well, do we want a decentralized or centralized system?" sometimes break down, because the answer is "for what?"

When the principles elucidating the Constitution serve us well for achieving whatever goal it is we want, then we're all strict Constitutionalists. When we're worried about the three-fifths compromise, the constitution is an unfit document that needs to be revised from time to time in order to improve it. The obvious example is Democrats love executive authority when there's a Democrat in the White House, and they love states' rights when there's a Republican in the White House. They hate states' rights when states are passing voter suppression laws, and they love states' rights when states are preserving Obamacare and mandating net neutrality, right?

But the way to solve that conundrum is not to say, "Well, you have an ideological incoherence as to whether you like states or the central government making your rule." The ideological incoherence is, "I want net neutrality, or universal health care, or socialism, or universal justice, and whatever tool in the moment serves that, is the tool I want to use." And my adversary can use it to weaken my goals. The reason we have the rules of the game is to produce the outcome that we want, not because we want to have a fun game, and maybe that's the difference between games as a kind of purely ludic exercise.

Although you sometimes get this with people who play pool like jerks, who just go all in on every hand in order to intimidate everyone else. Those people, if you're at the table—excuse me, not pool, poker—if you're at the table, you're like, "Dude, it's a friendly game and you're ruining it!" And they're like, "Well, it's in the rules." Everyone hates those people.

**R. GEHORSAM** Right. It seems to me that there are almost three domains that have different possible considerations for a social engineering piece. And one is kind of why we're here at InterPlanetary today, which is, how do we actually reenergize and galvanize society to value the whole question of us, as a species, becoming interplanetary? That's one set of activities, which I think is really interesting, and at hand.

The second one is sort of what I call the intermediate phase, which is the journey out. And there's where you get into the captain's rule, or the "lifeboat rules," where it's still very fragile, still exploratory. And then the third phase is, okay, things are more or less established. And now what are the relationships of those societies? I think it's interesting that we call them *colonies*; that has an implicit, well, colonial meaning. Let's just call them settlements, societies, something else, and what is their relationship to the mother ship now? That's really the complex, interesting, and probably conflict-ridden future.

**C. DOCTOROW** I think you skipped a step, because you started with "We all want to unite as a species to get off planet," but I think there are at least two camps. One says, "We all need to get off this planet before we use it up, because we can't possibly not use it all up. And if we use it all up in the process of getting off the planet, at least we got off the planet before we used it up," which has a certain circularity to it.

**R. GEHORSAM** Right.

**C. DOCTOROW** Right? And there's the other camp that says, "Well, we need to get off this planet so that we can learn to be better stewards of our own planet." And that eventual relationship is in large part dictated by why we go.

**R. GEHORSAM** Yes.

**J. FLACK** Let me push back a little bit. So two things are different about the future. One is what I said in the introduction, and that is that we have at our disposal potentially incredible data that we've just not had before, so we could have quantitative principles guiding our social engineering. And the other is that we will be going off-planet. These are two very big things, right? And so you guys are presenting a very coherent picture of the things that we might take into account, the steps we might pass through, but again, there are these questions: Are we really challenging ourselves? What are the limits of our imagination? Should we have that discussion? How can we identify what those limits might be? What are strategies we can use to reveal these biases?

**R. GEHORSAM** Good questions.

**J. FLACK** Put you on the spot.

**R. GEHORSAM** Well, I actually think this morning's panel, "Living in Space," was a good peek at some of the things that have to be considered. Antarctic missions where strangers are in close quarters is just one example. I think that a lot of rehearsal that can happen here, to suss out at least—

**J. FLACK** Counter-factual—

**R. GEHORSAM** Yeah. . . to suss out part of the things. The desire to . . . I forget how it was said exactly, but let's make sure we transmit the data back from these expeditions on Earth. Whether it's a SEAL team or whatever, we need to capture that data so that experiential knowledge can be transmitted in a meaningful and quantitative way. I think that's a way to start. I don't know if that gets at the question.

**J. FLACK** Well, what about video games?

**R. GEHORSAM** Video games are a way to approximate certain types of things, absolutely. I mean a famous, simple example that's roughly in this domain is the pandemic in *World of Warcraft*, where a virus was introduced, and behaviors could be observed. And so games are a very data-rich environment, and that's why I brought it up in the first place. Not so much because of the game play, but because of the data that's generated and available.

**D.A. WALLACH** I guess one of the questions you've raised, Jessica, is whether we should embrace the possibilities afforded to us by the sensorization of the world and our higher-fidelity ability to measure what's going on. And I think there is some debate among people who think about this stuff—maybe you're one of them—as to whether or not these sorts of systems are even in theory modelable, and interrogable, in an empirical way.

**J. FLACK** Absolutely.

**D.A. WALLACH** I'd like to bet that they are, because most other things that we've tried to measure have turned out to be.

**J. FLACK** Well, that's SFI.

We should have some sort of provisional holding period that we subject new knowledge to. When we think we've figured out how something works, we can start doing some fun sort of skunkworks projects with it, but we shouldn't immediately vote out every president in the world and ask for new people to implement the crazy, brand-new theory.

-233-

**D.A. WALLACH** That's SFI! I think what's important, though, and maybe missing particularly in the Silicon Valley world that I sort of inhabit, is a level of humility about how quickly and completely we learn and understand things, and what it does mean to understand something. We were talking before the panel about the history of science and politics that's littered with misunderstandings that we essentially overreacted to. And so we believe in things like phrenology, or other—

**C. DOCTOROW** Eugenics.

**D.A. WALLACH** —now silly ideas. Eugenics. And then we invest so deeply in them. We try to build our world around them. So we should have some sort of provisional holding period that we subject new knowledge to. When we think we've figured out how something works, we can start doing some fun sort of skunkworks projects with it, but we shouldn't immediately vote out every president in the world and ask for new people to implement the crazy, brand-new theory. That's inherently a bit conservative, but I think it's conservative in the right way. It's a question of what modesty looks like as a governing principle for our civilization.

**C. DOCTOROW** Can I propose that we actually do have a model for a civilization in which people have to live for a very long time, in close quarters, with limited contact with the outside, under very strict rules? It's prisons, right? And they're fully surveilled; we have tons of data about them. We're not very good at predicting how prisoners will behave. It turns out that human behavior is super adversarial. The people who want to influence, or control, or persuade people to behave in a certain way need to contend, first of all, with a second power block, or multiple power blocks, that want the same people to behave in a different way. "I want the voters to vote for me. You want the voters to vote for you." And there are the voters, who themselves are adapting and thinking, and have their own set of priorities.

One of the things that we know about persuasion and other forms of social engineering is that it's dynamic, and that tactics that work stale date really quickly. I was saying before the panel, if you go around Santa Fe you'll see on the sides of the old brick buildings old signs, ghost ads, "Buy Bar of Soap. Five Cents. It'll Make You Clean," and there was a time where apparently that sold soap, right? And now it's like, "AXE Body Spray Will Make You a Love God!"

LAUGHTER

**C. DOCTOROW** We've developed a callus over the part of our attention, or our persuadability, that, where "It Will Make You Clean" seemed like, "Well, they wouldn't say 'it will make you clean' if it wasn't true. They wouldn't let them put it on a building if it wasn't true, certainly."

LAUGHTER

**C. DOCTOROW** And now we're like, "Well, it's probably a lie." And so we see this with all persuasion techniques, even if we stipulate that Cambridge Analytica's marketing material should be taken at face value—that they are in fact Svengalis who can make you believe that up is down and black is white, and that they were making normal people into racists instead of just finding racists and making them into voters—even if you believe that that's the case, there's no reason to believe that they can do it again next year, right? Maybe it's a one-off.

**J. FLACK** I don't know if it's a one-off, but it's certainly early days.

-235-

**C. DOCTOROW** Right.

**J. FLACK** Okay, so maybe one more set of issues before we take a question or two. Is it naïve to think that in settling another planet we'll essentially have the opportunity to really start over? Is that naïve, or are there reasons to believe that these will be very isolated settlements, where there will be memory, of course, of human history and human institutions, but it will decay?

**R. GEHORSAM** I think it will decay over time. I think it really depends on how communication is preserved between Earth-based societies and other societies. And that's still very problematic and to be determined.

**C. DOCTOROW** I think that we wouldn't want it to not decay, inasmuch as we want it to be dynamic, because its circumstances would change, right? They're not LARPing day one of the colony for the next thousand years. If there are other people on other planets, and new technologies, and new social movements, and pandemics, and whatever, they're going to have to adapt. So I think that it's instructive to look to the project of framing the US Constitution, where you have these radicals who wanted to create a static system that would still flex a little. "We want to radically tear down everything; we want to set up a set of principles. They should never change, except very slowly and incrementally through these things like constitutional conventions." So they were pretty damn sure of themselves.

**J. FLACK** And they had a big impact, actually. Cultural history of the Constitution is very influential.

**C. DOCTOROW** Yeah, and it's a thing that a lot of people do when they constitutionalize. You see people trying with varied success to try and do it with free software projects or online communities. It almost never happens in games. Games have this problem, unless we're talking about an old-school MUD—they're super authoritarian in the same way that jokes are authoritarian. You don't get to vote when the punch line gets told. In a game, you don't get to vote when the dungeon master puts the orc in front of you; that's the dungeon master's job. Your job is to decide which dungeon master you want, and their job is to tell you where the orcs are. And so games are super anti-democratic in that way.

**J. FLACK** Hmm, interesting.

**D.A. WALLACH** I think it would be really dumb for us to start fresh. And the reason for that is that we've come up with a lot of cool stuff. And we also have a lot of evidence about the things that we can do.

**J. FLACK** I like the optimism.

**D.A. WALLACH** So you put us in certain circumstances, and unfortunately we can all, it seems, turn into Nazis. You put us in other circumstances, and it turns out we can sing "Kumbaya" and hug each other and worship the Rajneesh. And so I think we think of technology as "gizmos and gadgets," but it, of course, encompasses everything that we do and the ideas that we have in social norms, and those things serve to put us in a cage to ensure a certain amount of good behavior and limit bad behavior, but also hopefully not so restrictive a cage that we can't flourish as people or societies, or create new ideas. And I think that general framework of wanting to keep ourselves from being bad, but allowing ourselves to explore and invent, should also govern the way that we would move anywhere else. I think it would be terrible if we saw that as the beginning of history.

**J. FLACK** Okay, awesome. Maybe we have time for one question. Yeah, go ahead.

**AUDIENCE MEMBER** You discuss in the description of the panel capitalism and communism, and I wondered, well, communism workers say to bosses, "You pretend to pay us, we pretend to work." And then capitalism bosses say to workers, "All your innovations belong to us." Is there a system in which workers get out what they put in?

**C. DOCTOROW** So the question was—

**D.A. WALLACH** Do you want to rephrase it, Jessica, so that the audience can hear?

**C. DOCTOROW** Yeah.

**J. FLACK** So I think this is a criticism of both capitalism and communism, is that correct?

**AUDIENCE MEMBER:** Yeah.

**J. FLACK** And you want a better system, you want to . . . oh, let's wait for the train to go by.

TRAIN PASSES BEHIND THE STAGE

**J. FLACK** You want a better system in which workers get out what they put in, that maybe emphasizes the quality of effort more, is that right? Does anyone want to take that?

**C. DOCTOROW** So I think everybody wants to declare what they've done to be a unique act of creation, and all the things that they used to do it to be mere plumbing that anyone could have come up with. I don't know that you could ever calculate the part that the worker did, or the part that any one individual put in. The problem with measurement is that it pretends that it's measuring all the important things. And, generally, what it measures is the thing that the person who's doing the measuring is pretty good at, because that's the number that they want to make big and reify, when we're talking about performance measurement.

The reason we allocate resources at all is because we don't have enough. Having enough—we talked earlier this afternoon about scarcity and abundance—having enough is a matter of making things very well, which we're getting very good at, choosing what we want, which we're less good at, and then distributing what we have, which we're really, really bad at. It may be that if we fix one of those other two things, all of the questions about fair allocation go away. We don't worry about fair allocation of air, right? We don't worry about fair allocation of . . . well now we're worrying about fair allocation of sex. This is the kind of weird thing that happens when you start to complain, when you make allocation the big question. You start to say, "Well, why can't I buy a kidney? How come some people get to have sex and other people don't? Why can't rude, obnoxious, and objectionable people have sex?"

LAUGHTER

**C. DOCTOROW** There's now a terrorist movement grounded in this that just killed ten people in Toronto. So I don't know that allocation and fair shares are the right thing; maybe abundance is the right thing.

**D.A. WALLACH** I would just say, because we discussed it before the panel, Cory brought this up, I don't know if you were quoting him, but the philosopher Slavoj Žižek says this thing to illustrate for us how ideology is laid in our minds. He says that it's easier to envision the end of the Universe than the end of capitalism. And I guess when you think about these two systems opposed to each other, as you did in the question, what maybe we forget is that the future almost certainly is going to be characterized by new ideas that don't yet exist. And as a technologist, or as an investor, I'm always looking for the fountains of those new ideas.

I think at least for this crowd, since we're in America right now, it's important to recognize that China is producing some phenomenally interesting new ideas, some of which are terrifying and some of which are just sort of like, "Huh?" And they view themselves as instantiating socialism, quote unquote, "with a Chinese character."

Which, to Jessica's framing comments, involves a granularization of the way that workers are measured and how people's behaviors in the society are tracked and rewarded or punished. And so whether or not we want to bring that into reality, they are.

I think it's incumbent upon those of us in the West, or in traditions that prioritize the individual in a way that, say, China doesn't, to respond with equally new ideas that speak to deficiencies of our traditions, and offer a counterpoint to the proposals that they're making to the rest of society.

**J. FLACK** All right, excellent. Guys, thank you so much. Thank you. Thanks, guys.

# INTRODUCTION:
# VISUALIZATION &
# DESIGNING THE IMPOSSIBLE

No less than James Cameron, over scotch one Christmas dinner, said to me, "All technology is science fiction first."

Which at the time really irritated me.

You see, I had started out my career in high energy physics, and before that for a bit in medical physics, publishing papers while still an undergraduate— really hardcore. And, like a lot of science and technology super dorks, I had the firm idea that innovation and progress stemmed from "technological imagination."

And, therefore, that fictional technology was all just playful bullshit.

This was, however, a shockingly hypocritical viewpoint for me specifically to have, because shortly before I was to start my physics postdoc, the huge experiment I was going to join was canceled by the US Congress—hero to zero—and I needed a job. And I found one, writing video game engines.

Yeah. Suddenly, I was employed by a factory of fictional technology bullshit.

I spent the next couple of decades making software and hardware technology in the service of storytelling and world creation—in the service of making whatever fictional settings, worlds, and technologies that would produce the most delight and deepest emotional reaction from an audience.

Real technology in the service of fake technology.

But the thing was, we were actually making worlds full of devices and ideas that we *wished existed* in our world. That was why it was enthralling to us, and also to our audience. I was being my own, maximally ironic example of Jim's point.

Now, think about it. Your pocket computer, called a smartphone, also a video phone, and a moving map, and god knows what else. Space tourism (courtesy of a record company fortune!). Robots on other planets. 3D printers. AI doctors. Once you get going, nearly everything technologically cool we have around today was obviously yesterday's science fiction.

So, full circle. When technically gifted people dream, those dreams turn out to be set in the context of the fantasy worlds that inspired them in the first place. When the first cell phone team was dreaming, they saw *Star Trek* communicators. Generations later, we all carry them.

-241-

> Once you get going, nearly everything technologically cool we have around today was obviously yesterday's science fiction.

When we think of holograms, we see Princess Leia. When we see Boston Dynamics demos, we see the Terminator.

Ah yes. *Terminator*. Thanks, Jim!

Now let's see what some other people who have spent careers making fictional technology using real technology think about these ideas.

*—Seamus Blackley*
*CEO of Pacific Light & Hologram*

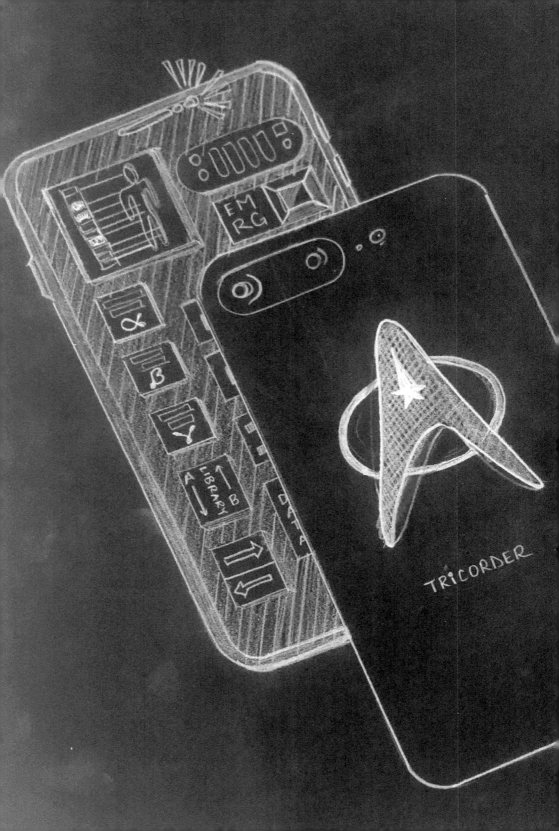

TRiCORDER

# VISUALIZATION & DESIGNING
## THE IMPOSSIBLE

*David Krakauer introduces the panel, moderated by
Seamus Blackley and featuring Doug Church, Scott Ross,
and Sasha Samochina.*

**DAVID KRAKAUER** Now let me introduce the panelists. Not
as cool as Mini-Van, but cool. [*Mini-Van is an affectionate nick-
name for Van Savage, the son of "Time Design" panel moderator Van
Savage, who was invited on stage to show off his space-tiger face paint
to the audience.*—Eds.]

I'll start with Seamus Blackley. Now, Seamus disappointed us at
InterPlanetary this year, because he was going to bring Richard
Feynman's van, which is covered in Feynman diagrams, for the
physics geeks out there. And to be fair, he tried, but it was too
precious, and he couldn't work it out. Driving that thing all the
way from California would have left Green's functions strewn
all over the highway.

LONG PAUSE. LAUGHTER.

**D. KRAKAUER** That was a physics joke! Anyway, Seamus
Blackley, also known as the Father of the Xbox, for those of you
who know about games consoles, here he is, the designer. He
is trained as a high energy physicist, he's a game designer, and
he's also the CEO of the venture-backed start-up Pacific Light &
Hologram. That's Seamus.

I'm doing this slightly out of order. Now, Sasha, fantastic! Sasha
Samochina is an immersive visualization producer at NASA's

Jet Propulsion Laboratory in Pasadena, California. She actually joined that team after working in New York in video and web development. But, before that, she was head of media and a producer at the Field Museum of Natural History in Chicago, which is an incredible institution and something that we've all really appreciated at SFI.

Sitting next to her is an old friend, Doug Church. I like this, this says, "Doug Church is an American." That's great! Okay, it actually says, "American video game designer." Anyway, the fact is, he is one of the pioneers and visionaries of CGI and 3D gaming. He went to work at Looking Glass Studios. He worked in *Sim* games. You probably know him from *Ultima Underworld*, *System Shock*, and *Thief*. And, he is one of the people who coined the phrase *immersive simulation*.

Last but not least, sitting at the end there, an old buddy, a face that we know, Scott Ross. Scott Ross partnered with George Lucas in a number of his companies. He founded Digital Domain with James Cameron, and he is now also doing all sorts of amazing work in VR. So welcome to the panel.

**PANEL** Thank you.

APPLAUSE

**SEAMUS BLACKLEY** All right, thank you, David. So we're here to talk about visualizing the impossible, which is a semi-impossible thing to talk about. And what we really mean by this is that throughout our lives we're bombarded, and since the dawn of media we've been bombarded, with visions of the future. People as far back as the time of the Sumerians would write about the future, would draw about the future, would create imagery about the future. And that imagery and those ideas, which have been presented to us throughout the centuries, have been very influential in the development of technology itself. More often than not, you'll find that people who developed technology, when asked about it, will tell you that they had in their mind something that they read as a kid, or something that they saw in a television show or in a movie.

And so one of the things this outstanding group is going to talk about, what I'm going to ask them now is, do we really believe in the influence of imagery? And if we do, how powerful is the influence? How do we go about picking out those brains, those minds who are going to invent the future, and how do we focus those minds on the things that are most important to develop? How much power do we have in that? So I'll throw it over to the guy at the end, Scott, our movie guy. Go, Movie Guy!

**SCOTT ROSS** Well, you know over the years working on screenplays, it was rare, if ever, that the technology people actually had a direct input on the vision of the director or the writer. Almost always, the vision of the director and the writer came before any sort of technique. So oftentimes technology was invented as a result. For example, in *The Abyss,* with the water weenie creature, with Mary Elizabeth Mastrantonio, that whole thing was something that was written by Jim Cameron, and we had no idea how to do it. And so it was the creative that actually drove the technology.

**S. BLACKLEY** Do you agree with that, Doug?

**DOUG CHURCH** Yeah, I think I generally do. And, you know, Cory Doctorow on the previous panel mentioned "shooting at the barn and then going and circling the thing you hit," and I think that is the model for how a lot of this sometimes works: the creative gives the vision, and then investigation happens. And then you go back and say, "Oh, that's what we were doing all along. Yeah, that's awesome." So I feel like that loop is how it happens often.

**S. BLACKLEY** When you say that, though, it's a special thing because something that many people may not realize about game development is that you're holding both the creative vision and the technology at the same time. And they are having to inform one another, and it can either shackle the creative process or it can sort of unleash it and get you into a lot of trouble, like the trouble we were in at Looking Glass together [*speaking to Doug Church*] by trying to do things that hadn't been done before and then trying to figure out how to back-build technology into it.

And I think, Sasha, this is almost your job description at JPL, right?

**SASHA SAMOCHINA** Yeah. I'm sitting quietly and patiently thinking, How do I phrase this perfectly? So what I do, quite literally, is make things. I create projects to help scientists and engineers do their jobs. So I create software that's not flight-critical for NASA, but it's a tool that will help scientists and engineers do what they do, to make their lives a little easier.

One example is a Mars touring reconstruction. So we have planetary geologists exploring the surface of Mars through the HoloLens device that Microsoft makes, looking at actual rock formations and doing real science. Translating that to the public is quite hard to do unless you have this technology in the palm of your hand, so part of my job is to take that and do public outreach and to make more web-accessible things. We have a web version of this called "Access Mars." But does that tell enough of a story? Is someone going to look at Access Mars and say, "Oh, it's Mars, I'm actually seeing something." You need the narrative, right? Well, the narrative is that scientists are using it every day.

And I talked to a former JPL engineer, and he asked, "What's the first and most important thing that happens at JPL in any mission formulation?" He worked there twenty years ago, and I said, "I don't know." That's a great answer that we constantly use at JPL; I encourage everyone to say "I don't know" if you don't actually know. Anyway, he said, "The most important part of formulation is the artistic concept that comes before a mission is brought to fruition, so that people can visualize what they're actually talking about." A lot of that comes from actual data that we already have, and some of it is projected data that we might have. So I guess my answer is kind of on the nose.

**S. BLACKLEY** Your job is predicated upon the belief, your existence is predicated on the belief, that the most powerful thing that we can do to inspire the future we want is to imagine it, and distribute that imagination everywhere we can possibly think of.

**S. SAMOCHINA** Yeah, if anything it's just me screaming, "We have all this stuff and you should use it!" And, you know, when people see the visualization of Mars for the first time, they think it's just a video game. I mean, we do build it in the Unity engine, and it is a visualization, so we take data from the Curiosity rover—the photos that Curiosity takes on Mars—and we mix it with orbital data, and then create a pipeline that is dynamic and updates every time there's a downlink from the Curiosity rover. When you explain that part of it to people, they say, "Oh my gosh, that's amazing!" Otherwise, they're like, "Am I just in New Mexico?" Because it looks like you're standing in the desert.

-247-

**S. BLACKLEY** Well, I have to say, as a guy from New Mexico who makes video games, both of those are compliments: that it looks like a game and that you could be in New Mexico.

LAUGHTER

**S. BLACKLEY** So I'm all thumbs-up! But it's interesting because you're intentionally doing this: you're intentionally taking scientific progress and wrapping it up in media in order to inspire people to believe in it, to be excited about it, and to get funding for it, which is very important to do. Scott, we're unintentionally doing this. You're hanging out with Jimbo Cameron, and he's yelling and screaming and nailing shit to the wall. And you're scrambling around trying to figure out how to deal with this vision. Unintentionally, or semi-intentionally at least compared to Sasha, you're projecting a vision of the future and a technology of the future in a very compelling narrative that really has a lot of influence.

While you were doing that, did this occur to you?

**S. ROSS** Absolutely not, no, without a doubt. Your head is so deep into thinking, I've got a release date that's on June the fourth.

It's this film, it's going to be put out in five thousand theaters," etc. You're not thinking about where it might go or what it might do. You're just trying to solve a problem that's in front of you.

**S. BLACKLEY** Can you think of places in your own work, or in other movies that you love and respect, where we've seen the birth of a new technology in media? Something we can see around us today.

**S. ROSS** Well, computer graphics, as a whole. I go back to the point in time when I was running Industrial Light & Magic, and all of the work was done photochemically. There was very little, if any, computer-generated imagery at the time. And when computer-generated imagery first came about with the Pixar cube [image computer]—before Pixar, the Pixar cube—and we started doing things for *Star Trek* or movies like that, most of the entire company pooh-poohed it, and said that it was never going to go anywhere. So the adaptation of computer-generated imagery really has changed the way feature film and so many other things like video games are made, even beyond entertainment. It's changed the way we look at the world.

**S. BLACKLEY** Well, I think that this segues directly to Doug. I spent possibly two years or more in a small smelly room with Doug trying to code up—

LAUGHTER

**D. CHURCH** Software development. *Shrugs.*

**S. BLACKLEY** —some of the first-ever, real-time 3D graphics and real-time physics systems. And I can remember, I think we were at Logan Airport picking somebody up, and we were drawing tiles on the floor to figure out sorting. We didn't even really know to call it *sorting* because it wasn't something you could look up. So you've been right at the center of the development of that computer graphics technology into interactive technology. And now, for instance in Sasha's case with NASA, that is being used as a primary scientific instrument.

**D. CHURCH** Yeah, and even going one step further, think about *Star Trek*-type movie visualization. Think about the Tricorder. There was a visualization of a lot of things we were going to be able to do with this amazing device. We then ended up building pieces of glass which run our lives. It's not clear if that's good or bad, and it's not the same thing, but it came from that inspiration. A bunch of people in this space took that high concept and went with it. I think you see that in a lot of these mediums where you find some piece of an idea, and you focus in on it. I want to represent flying, and I want to build a flight simulator. How do I do that? And what are the pieces? That's a random hypothetical.

-249-

**S. BLACKLEY** Random, yeah. *Laughs.*

**D. CHURCH** And, you know, you end up doing something very different than what you intended, but sort of along the same lines of the original inspiration. So it's not that you need the inspirations because they're the product feed or the spec, you need the inspiration to start digging into the detail, to start discovering what's actually worth doing.

**S. BLACKLEY** Right. And so the process of imagining the future and then having to implement it technically—

**D. CHURCH** And realizing that you can't do the thing you imagined—

**S. SAMOCHINA** And the tricorder is a great example because the people that are building technology that's like the tricorder, especially at JPL, were watching *Star Trek,* and they didn't even realize that they wanted to make this thing, until they made the thing and were like, "Wait, isn't this the . . ."

**S. BLACKLEY** Then when you're presented with a problem, you have in your mind already a map toward where you might want to go. I think you might think of the piece of glass that rules your life—if you have children you might think of it as the piece of glass that is also a babysitter. It's many things to many people, good and bad. But the cellular technology on which that was built was imagined by people who probably wanted a *Star Trek* communicator. The satellite systems that had to be in place to handle the data of GPS were built by people who had a different idea, but I think the power of the imagery that we create for entertainment and storytelling is the power that sets a road map in people's minds, so that when they sit down to go to work on something, they're not lost. There's a path to follow.

**S. ROSS** I believe you have to have the context through which to imagine something and, if this context doesn't exist it's very difficult to imagine what could be. For example, you know, how many of us really remember what it was like being six months old? Or nine months old? A year old? Some people say they do, but we didn't have the context of language to wrap those ideas with. And we really started to understand what life was, where we could describe it as language skills came in.

So imagining the future, I think, has to do with the context of imagining what could be from an image or a storytelling, or a creative point of view.

**S. SAMOCHINA** Right, and tying that future into the exoplanet work that we do at NASA, I mean, our favorite stories to write for the public are the stories that connect them to something like *Star Wars*, which is way more tangible for people than talking about planets and stars way outside of our Solar System. This kind of science communication realm really helps tell stories.

**S. BLACKLEY** This begs a larger issue. So it's been fun to talk about, but I think we all fundamentally feel, and maybe everybody here does, that the ways we imagine and talk about the future—the stories we tell about the future—influence the direction of our future in a very strong way. So one of the things that's then incumbent upon us in understanding that we have this powerful tool is to tell the right stories that move the human race in the right direction. Not only towards survival, but towards a future that's meaningful.

So in that regard, what do you think are some of the directions we should try to push media into?

**S. ROSS** One of the things we were talking about before the panel that I'd like to bring up before our time runs out is the whole concept of machine learning and artificial intelligence and what that might do to designing, developing, and contextualizing what creativity is and, therefore, what technology might be.

..................................................................................................

As someone said on the previous panel, if we're talking about the future when we're here sitting in North America, it's hard to picture a world without capitalism. We're also pretty used to gravity.

..................................................................................

**S. BLACKLEY** Yeah, for sure.

**S. ROSS** Anybody have any ideas on that?

**D. CHURCH** As someone said on the previous panel, if we're talking about the future when we're here sitting in North America, it's hard to picture a world without capitalism. We're also pretty used to gravity.

LAUGHTER

**D. CHURCH** We're pretty used to a bunch of annoying stuff. That's how it is. If we're envisioning this future, how do we put ourselves in a place where those concepts or stories are surprises to us as the people who are creating them? And what are the things we can do to augment our creative capacity? Or surprise us into other ideas? Or present us with possibilities we haven't considered, so that we can stop iterating? Once again, someone on the previous panel mentioned that we don't want to throw it all out, and I certainly understand that, but how do we push ourselves out of the comfort if we're envisioning a future? How do we start picturing new things?

**S. BLACKLEY** Well, when we talk about space travel and we talk about the concepts that we're dealing with here at the InterPlanetary Festival, one of the things that always occurs to me is the horrifying limited lifespan of human beings. And, you know, we do throw it all away! We learn all of this stuff, we spend half our lives learning, and hopefully we'll make some contribution. How do we ensure those contributions get carried forward?

. . . there's a lot of space in space . . .

It's difficult to do in a specific way. It's difficult for me to pass forward 650 pages of equations that took me half my life to figure out. You can pass a spear, you can pass a set of ideas, you can pass a direction, and that's something that visual media and storytelling media are so powerful at.

**S. SAMOCHINA** I think my favorite version of passing along knowledge of our existence is Voyager and the Golden Record. I think that's a pretty good way to do it. It makes us still feel like there is something out there showing that we even existed, so I'm glad that happened when Voyager launched. Or both Voyagers, as it were. But I wish there was more of that happening, metaphorically.

**S. BLACKLEY** If everything else goes to hell, we have the Golden Record, man.

**S. SAMOCHINA** The Golden Record is out there, and maybe no one will ever find it because there's a lot of space in space, but at least it's out there.

**S. BLACKLEY** Yeah, there's a terrific essay from Stephen Wolfram who is helping a friend to make quartz disks like that. And the way he characterizes this product is not as communicating the progress of the human race throughout the galaxy. Rather, it's to create a monument in case we vanish. Such a sad motivation.

**D. CHURCH** Yeah.

**S. SAMOCHINA** Yeah, well, it's good, too. I'm a serial optimist, that's all I have to say. But I think that the transfer of knowledge is interesting and making that more of a machine-learning thing is complex because that's trying to make machines think like us. Can we let go enough? There are all these questions and—I mean, I'm all about it—but as we all know, it has to be done the correct way.   -253-

**S. BLACKLEY** Well, you exist in a possibly freer space than all the people making the decisions. I'm going to keep on coming back to this idea of responsibility about direction. When we're making video games, the primary thing on our mind is the idea. We're super excited about this idea. But also, how am I going to get funded for this? How is this going to work commercially? And that removes or sort of truncates, possibly savagely, the creative space in which we can think about trying to pass positive messages forward.

Recently, Neal Stephenson, who was speaking earlier, participated in a program with a bunch of writers to talk about positive science fiction for the future, nondystopian science fiction books. And I believe it's true that *Seveneves* grew out of this. I think that in all media, this is the thing that should be on our minds despite all of the constraints that are always placed on these processes.

...............................................................................

## If everything else goes to hell, we have the Golden Record, man.

...............................................................................

**S. ROSS** And the things that sell are generally dystopian stories, more so than utopian stories. A little pitch: I'm introducing *Terminator 2* tonight, which is the ultimate dystopian story, right? Last night we screened *The Fifth Element*, which is sort of a utopian story. But it seems like the David Finchers of the world continue to push that dystopian sensibility out there, and I don't know how we address that. I think it's our moral responsibility to, but how does one?

**S. BLACKLEY** Well, it is, because that story is caught up with everything else when you go to develop technology. I think you mentioned machine learning in artificial intelligence, and all of those stories are constantly on the minds of the people developing artificial intelligence. They think it's preposterous. I read media stories saying, "Oh my god, why are people developing intelligent machines? Don't they know they're going kill us?" You know the guys working on these things are not frickin' idiots. They've seen *Terminator*! They understand the danger, and they're thinking about it all the time.

**S. SAMOCHINA** Well, we hope they've seen *Terminator*.

**S. BLACKLEY** Yeah, well, we hope. And if they haven't, I'd like to make it a law! But there's a real responsibility to ensure that the mixture of future road maps is rich enough.

**D. CHURCH** The tone is harder to pass on than the artifacts. You know, we can build pieces of conceptual work, whether it's a piece of art or a piece of glass, and pass that on, but the overall tone is harder to change and speaks to a lot of the social engineering stuff from the previous panel, about how and where do we aim that.

**S. BLACKLEY** Right.

**D. CHURCH** Because that's much harder to do—you can't just build that and hand it off.

**S. BLACKLEY** True.

**S. SAMOCHINA** I'm also of the school of humanizing the robotics in a way. A friend of mine who just graduated from MIT wrote a thesis about her idea, which was about prototyping prototypes, or building robots that build robots. Basically, you would never have to buy a fork again because you can 3D-print one. It's a practical way of not making waste happen, of connecting to other people, and of making things out of really easy parts. It's about making things a little bit easier, and less about building a big bad robot.

**S. BLACKLEY** Yeah, I think it's a thought process. Now, we're not going to decide here, nor is any group of people for any finite amount of time going to decide what this direction should be, or how to handle the problems. But I think it's important to think about. It needs to factor in and, and we need to understand the power of that.

You know, I'm sure if Gene Roddenberry were with us today, he would be thinking and talking a lot about the legacy of the storytelling and the technology that followed in the wake of *Star Trek*, and how important that was. And how lucky it was that he presented such a positive vision. If the television show that sat in the place of *Star Trek* had been dystopian, I am sure that the world would be a worse place right now. It's really something to think about.                -255-

**S. ROSS** And it gets even more powerful in the world of mixed reality and VR because now we have neural plastic tissues, and things really get much more real and much more difficult in telling those stories.

**S. BLACKLEY** Yeah, Doug, how do you see the power of story-telling changing as we move into a mixed-reality future? Small question.

**D. CHURCH** Yeah.

LAUGHTER

**S. BLACKLEY** Give it thirty seconds. No big deal.

**D. CHURCH** I do think there's a degree of presence and immediacy that fools—maybe fools is the wrong word—*engages* the brain more than it has before. We've taken away a bunch of levels of distraction. We've taken away a bunch of indirect actions, and we've replaced them with direct actions like reaching out and shaking that person's hand or patting them on the shoulder. That's a very different experience than hitting the A button and having the text "They smile at you" pop up.

LAUGHTER

**S. SAMOCHINA** Unless you're into that!

**S. BLACKLEY** Thank god you don't write video games!

**D. CHURCH** Totally fair. But there is an immediate lizard brain thing happening as we drive more towards that fidelity and that connection, and so the responsibility goes up. But we have a bunch of creators who are maybe not in that mind-set yet, and so where does that all land as a challenge?

**S. BLACKLEY** It sounds like what you're saying is that the peril of that responsibility only increases as our storytelling power increases, right?

**S. SAMOCHINA** Challenge accepted!

**S. BLACKLEY** Yeah?

**S. SAMOCHINA** I guess so, yeah.

**S. BLACKLEY** So we're almost done, but where would you like to push this? Where do you want to see this go, Scott?

**S. ROSS** Well, again, I think we get back to that issue that words are powerful. And that if you envision something, then you have the possibility of having it come true. So if that's a reality—and I believe that it is—I think we have to start to look towards building utopian worlds and building a future that is much more loving. I think that's critical.

**D. CHURCH** I would say increased focus on envisioning the systems, and the context, and the interactions, and less focus on the pure artifacts. More focus on the context for interaction between people and how that goes forward. Because I think it's going to be super important if you go to space—frankly, it's super important already for all the problems we're having right now—but I'd love to see us focus a little more there, even though it's harder.

**S. SAMOCHINA** For me, I think that the technology is growing. I work in the field where I'm making something kind of specific that I try to communicate with everyone else. But what I stress to everyone, always, is that I want the audience to become the creators. I want you all to see the content that NASA puts out because it's there for you to use.

**S. BLACKLEY** And you paid for it!

**S. SAMOCHINA** There is so much. I haven't even seen it all, and I will never see it all. It's infinite and ever expanding, sort of like space. And it will always continue. I'm an artist who works in science. I want to wave the flag and say that that is a possibility, especially to young kids who are being asked which school they're going to choose. I say choose both and carve your own path out of it. And I think that anything is possible in the realm of believing in the content and creating more stuff. So keep making stuff!

**S. BLACKLEY** Definitely. Definitely keep making stuff! Well, thank you guys very much. Thank you to the panel. Another excellent discussion.

# INTRODUCTION:
## THE END OF THE WORLD?

When trying to think about complex and speculative ideas, it's usually a good idea to take a step back and ask: what are some things we know, what are the things we can't know, and what are some areas we can make guesses around?

I think that, as long as people are around, there will be love.

Let me explain. As we know from the second law of thermodynamics—also known as entropy—there are many more paths to disorder than there are paths to order. Surely, over time, disorder and dispersion will dominate. We would be wise to take heed and understand that there is a real possibility of sliding back.

As I see it, love is about seeing and creating beauty. Whereas love restores beauty, entropy destroys with no purpose. I submit to you that if we are to survive for any length, love has to be our guiding star for navigating through the fabric of the physical world.

How about things we can guess?

When I think about the future, I often imagine what it would be like to be a Starfleet officer on the bridge of the *USS Enterprise* in the twenty-fourth century. I wonder about what it would be like to explore deep space and to encounter entirely new civilizations, discovering new places, and studying physical processes that are beyond our current awareness. I also hope that the crew is happy! I think it would be thoroughly sad if the daring explorers of the future live with the same mindset and the same type of standards as we do in the twenty-first century. How disappointing it would be if the explorers of tomorrow travel at many times the speed of light and still become irritated and annoyed, forget to be civil to each other, and shout and threaten as means of self-expression and communication.

Our brains are designed to keep us safe in the wild, not to make us happy. But as we continue to master the physical world around us, life-threatening scenarios of the past such as drought, famine, and physical threats are no longer viable threats. Instead, we've created technologies and social structures to enable each person to have more and more power. Each of us now has more power and resources at our disposal than ever before. This power can be used for good, for destruction, for nothing at all. So if we all become so powerful as individuals, doesn't it make sense to have an upgrade in our thinking and information processing, so that we can not only make more positive contributions but also live more enjoyable and fulfilling lives?

-259-

> If we are to survive for any length, love has to be our guiding star for navigating through the fabric of the physical world.

To date, much of psychology has been focused on helping individuals in pain reach a more manageable level of suffering. While this is an important and humane pursuit, why not also focus on understanding how to create more fulfilling, rich, and meaningful lives too? Only a small amount of resources has been put on this pursuit, but my hope for the future is that we all become aware of our own individual potentials and sense of possibilities. Perhaps living with a higher quality of emotions and experiences may be the most valuable achievement for humanity.

Looking at the near to midterm, I think we can make guesses about some areas of likely technology-linked developments. Applications of artificial intelligence and robotics in transportation, manufacturing, food industry, and elderly care are likely to have a major impact in the near term. Similarly, commercial spaceflight, computing, 3D printing, communications, and new financial tools will continue to have impacts on commerce and governance.

The development of lab-grown meats (a.k.a. cell-grown meat, or clean meat) is an advancement only few years away with the potential to radically impact traditional farming and land use, improve nutrition in all parts of the world, and make major positive environmental impacts. This is one area that I'm particularly excited about and consider as a major step forward for all humanity.

Looking farther out, advances in medicine, especially genomics, will likely lead to their natural apex of longevity, of massive life extension. Serious longevity is still many years and innovations away, but the impact will be felt in all aspects of life ranging from real estate and taxation to governance and scientific advances. Consider, for example, that currently it takes roughly twenty-five to thirty years for a professional scientist to master their craft. They will likely have only forty years of real impact on their field before retirement. Longevity can result in astonishing advances in all areas of sciences.

How about the things we can't even imagine today?

We humans have shown that we're severely disadvantaged at recognizing our own cognitive biases, and similarly fail at intuitively understanding nonlinear and multivariable concepts. In short, we like the world to be simple, linear, and repeatable.

We desperately need tools to advance our cognition and decision making. Will these tools ultimately take the form of brain implants? In a world of ubiquitous sensors and AI, will we want to make any decisions at all? Will we want to concentrate on being fulfilled? Will we want to eventually merge with the machines we've created and become dematerialized entirely? It's hard to tell, but we must continue thinking about these long-term strategies for what it would mean to be a human and design our path forward. After all, technology should be developed to be at our service, and history, in many ways, is ours to write. And perhaps we should start by being less offended, taking the time to be even more curious and appreciate with awe the world around us, and pursuing our passions without taking ourselves too seriously. Most importantly, bring love into all our actions, now.

—*Armin Ellis*
*Founder, Exploration Institute*

# PANEL:
# END OF THE WORLD

*David Krakauer introduces the panel, moderated by Annalee Newitz and featuring Armin Ellis and Lauren Oliver.*

**DAVID KRAKAUER** Hello? Hello? Oh good, we're on. Okay. This is the final InterPlanetary panel. And at first, you know, when we were envisaging all this, we thought we'd go out with a dystopian bang, but now we've realized that the vibe of this festival is so positive we have to turn it all around, and the panel has to talk not about the end of the world, but the future of the world and how we are going to save it. That's important!

So let me just do a brief introduction to these incredible people. It'll be really brief. Here's Annalee Newitz on my left, who's been to Santa Fe before to give a public lecture for the Santa Fe Institute that went incredibly well. She recently published a novel, *Autonomous*, that was released by the publisher, Tor. She's also the author—and this is partly why we thought of you—of the book *Scatter, Adapt, and Remember: How Humans Will Survive a Mass Extinction*, which I guess is kind of positive, in a way . . .

**ANNALEE NEWITZ** It has a happy ending.

**D. KRAKAUER** She's also the founding editor of one of the best webpages that was ever made, i09, if you're the sci-fi type. And she's the editor-in-chief of Gizmodo, and lots and lots of other things.

Armin Ellis, on that end, is a scientist and an engineer. He worked on some pretty hard-core infrastructure for space missions at the Jet Propulsion Laboratory; in fact, he worked with Pete Worden, who you might have heard yesterday. But he has taken the insights that he gained as an engineer working in these space projects to think about how we think about making society function more efficiently. And I think you'll hear a lot about that today. He's doing amazing work in that area.

Lauren Oliver is a person I've known forever. Written here it says, "Never provided a bio," and that's it. But, fortunately, I know Lauren, so she doesn't escape me. On her webpage, which you should all go and see, called Quellette Studio, this is how she describes herself: "Artistic type, obscurist, powder monkey." But she's also an amazing artist, and we showed for a number of months up at the Institute her work where she reenvisages the poles as a colony of occupation, invasion, rediscovery, with some dystopian environmental calamity, and it's the most beautiful poetic photographic exhibition of a world that doesn't exist that I've seen in ages. So she's extremely talented. And with that, panel!

> One of the biggest dangers to creative thinking—and therefore getting ourselves to solve problems of the caliber that we're talking about—is really a lack of independent thinking.

**A. NEWITZ** Thanks very much.

**LAUREN OLIVER** Thank you.

**A. NEWITZ** As David said, we were given this panel topic titled "The End of the World?"—question mark. And when we spoke to each other about it, we said, "Well, no!" So we've subverted the panel, and we're going to be talking about saving the world. But before we get to that, I want to start by talking a little bit about what you guys think are some of the biggest problems that humanity is facing right now. Before we get to the good part, the happy ending, what are some of those problems? And what about your work brought you to those problems? Lauren, why don't you start?

**L. OLIVER** The thing that scares me the most is that the oceans provide so much of our oxygen, and we're killing them. Maybe Armin can explain exactly how. But because they absorb so much *climate*, they're producing less oxygen, and it's my understanding that we need to breathe oxygen. It's sort of fundamental to our existence. So this idea of all of us starting to slowly suffocate, that's what keeps me up at night.

**A. NEWITZ** What do you think, Armin?

**ARMIN ELLIS** Well, just by way of background, I've had the opportunity to work on some very, very complex projects. A lot of space missions require you to solve problems in ways that haven't really been solved before. Now I'm applying a lot of these processes to different organizations, different kinds of projects for nonprofits even, and individuals. One of the things that I've noticed is that one of the biggest dangers to creative thinking—and therefore getting ourselves to solve problems of the caliber that we're talking about— is really a lack of independent thinking. We're into group thinking, and thereby we're preventing ourselves from coming up with some major breakthrough ideas. I see that as a major threat right now.

**A. NEWITZ** I'm a science journalist, but I also write science fiction, so I've had to think a lot about how to tell the truth about what's happening on the planet, scientifically. But then I also get to lie about it in my fiction, and I get to be very opinionated about it. One of the things that I think is great about science fiction—at least the kind of sci-fi that I write, which is very much about science and scientists—is that it lets us tell stories about the ethics and the culture surrounding the questions that can be answered using scientific instruments. One of the problems that I've encountered over and over again, one that I feel we're still trying to grapple with, is the aftermath of colonialism all across the globe. Here in the States, yes, but also in Asia, and in Europe.

I say that because colonialism doesn't just create ongoing economic problems, ongoing problems between nations, or between ethnic groups. It has also changed the way that we treat habitats, and it's changed the way we treat the environment. Colonialism has led to a lot of problems that today we would consider the task of environmental science to handle, and it's led to all kinds of lingering social problems around everything from slavery to occupation of land. That's one of the things that I think about, those interconnected social and scientific problems. These are huge problems. Some of these problems that are scientific are problems that have to do with the human mind, and also to do with history and society.

Now let's get to the tough part. I think that it's tempting, especially when we're talking about things like the environment, to get paralyzed, to feel like we've just totally screwed up, that there's nothing we can do, that we just need to be mourning the loss of the planet, the loss of humanity. And we just go, "Screw it, there's nothing we can do!" How do we move on from feeling terrible about mistakes that we've made to a mindset where we can start to fix them? Armin, I know you work with groups that are trying to overcome problems.

**A. ELLIS** Yes.

**A. NEWITZ** How do you approach that?

**A. ELLIS** I mean, the first thing is to just stop.

**A. NEWITZ** Stop it!

**A. ELLIS** Just don't make things worse than they are. Really, the first thing to do is to just sit back and relax for a moment so that you can actually start to get a different kind of perspective on the problem. When you're scared, good ideas aren't going to flow. How about sitting back, looking at the situation as is—not the worst-case scenario, but as it is right now—and then starting to work the problem? Once you get started with this kind of a process, it's really important to feel safe and to start to play with all sorts of ideas. Start to throw around concepts, ideas that may go nowhere. You may be dead wrong, and it might not be what everyone else thinks, but that's okay. That's a part of the process. Creativity is a messy business, and we should be okay with that.

**A. NEWITZ** I'm curious because I know you have a background in science and engineering.

**A. ELLIS** Yes.

-267-

**A. NEWITZ** And what you're describing does not sound very much like a scientific process, does it? Is there a connection between your scientific background and how you're solving problems now in your job?

**A. ELLIS** Well, there are a lot of scientists who solve problems in a very linear way, and there are a lot of artists who approach their own art in a very linear way. Then you have artists who solve things in a radical and new way. The same thing happens in the sciences. Actually, I would very much say that science and engineering are far more creative human pursuits than we tend to give them credit for. Scientists and engineers and our approach to it, our personalities, really do make a massive difference in terms of the solutions that we end up coming to at the end of it. What do you guys think?

**L. OLIVER** I loved what you said earlier today about collaboration, about facilitating collaboration, and taking the ego out of the solutions. I think that so many of us talk about labels. People are in this camp. People are in that camp. The problems that we're facing are so huge, they're global, so we absolutely have to put away our differences and look at our basic survival instinct and come together. It sounds kind of *kumbaya*, but literally we have to get into survival mode.

**A. NEWITZ** Lauren, can you just talk a little bit about your art project and how you've thought about bringing a message about the environment forward through writing about space owls and creating space owls?

**L. OLIVER** Okay. The art project that I'm working on, I've been working on for a few years, it's called *Ice-Station Quellette*. It was over at Phil Space, then it was up at SFI. Now it's at Meow Wolf. If you've been over there, you probably know it better as the big furry, tall, blue thing with the blinking eyes and the horns.

**A. NEWITZ** How many people in the audience have been to Meow Wolf?

APPLAUSE

**AUDIENCE MEMBER** Go tonight!

*The first InterPlanetary Festival closed with a performance by computational biologist and renowned VJ/DJ Max Cooper. The show was co-presented with Meow Wolf, and Max would be playing a special after-event set at Meow Wolf for InterPlanetary attendees.*

**A. NEWITZ** Yes. Go tonight.

**L. OLIVER** Go tonight! It's going to be fun. So the space owl is the central figure in a little science fiction story that I'm working on now. I'm working on expanding it. I'm really interested in reaching young people and children of all ages, because I'm trying to create an emotional connection between the story—the thing that's going on—and the science fiction, which really is the backdrop for our reality right now. I mean, the stakes are really high and we're in a pivotal position. It seems like the best way to reach people is to create an emotional connection so that you can feel it in your own life. And sometimes a story is the way that we relate things to our own life; we become involved with characters and in the story itself that may have parallels in our own life. My story is pretty simple. The bad guys will do anything to win, and the bad guys in my story are called "monopolists." They are the fossil fuel industry that's been concealing the greenhouse effect, which they've known about for half a century, as have our leaders.

Instead of using that lead time to do anything, they've led us down the primrose path where it's like, "Burn it, burn it! We'll be okay! Somebody will figure something out." We were talking about colonialism, and I think we live in a society where we have outsourced responsibility to others, and we think there's going to be a technological fix. My story is about encouraging people to take direct action in their own lives, if only on an evolutionary level that's completely personal.

**A. NEWITZ** Yeah. The thing that I love about writing fiction— aside from the fact that I get to actually lie, and I don't have to check my sources, unlike with journalism . . . hopefully—is that you can allow yourself to imagine scenarios that you might not allow yourself to imagine if you were writing a work of straight fact. Because facts have a way of hemming us in. I mean, they have to, partly, because that's the balance of reality. But they also make us sad, and they make us feel helpless. For example, my recent novel, *Autonomous*, is about an artificial intelligence in a world where humans have decided to put robots of basically human-equivalent intelligence in indentured servitude for the first ten years of their lives in order to pay off the cost of their manufacture. When you think about it economically, it makes perfect sense.

But once you start empathizing with a character who's in that position, who happens to be a robot but is also born into indenture and can't control his own thoughts or his own feelings, I think there's a way that it allows us to think about and empathize with people in that situation. It allows us to think about the implications of research that's going on right now. I've talked to several AI researchers who say only half jokingly that they want to create happy slaves. If you are creating a generalized intelligence, the definition of a generalized intelligence is that it will not be happy being a slave. As you're in the moment trying to create that thing, it's hard to think, to extrapolate, and say to yourself, "Wait, what's going to be happening with all of these sapient beings that we're making and putting into this role of a new slave class one hundred years from now?"

I think that you always have to have storytelling happening right alongside science, because storytelling is where we ask those ethical questions that sometimes sound silly, but we need that space to ask those questions. It's kind of like the safe space to be wrong. You need to be able to just imagine what's going to happen and then look at the consequences before you build the thing that you hope is going to be a happy slave. Otherwise, maybe you deprive the oceans of oxygen.

Okay. We're thinking about ways to save the world, and we've talked a little bit about how to overcome that sense of paralysis. So now I want to ask you guys some fun questions. I mean, we are here at the InterPlanetary Festival, are we going to solve some of these problems by going to space?

**L. OLIVER** Take it away.

**A. NEWITZ** Yeah, take it away.

**A. ELLIS** Well, I have some opinions about that.

**A. NEWITZ** I thought you might!

**A. ELLIS** I happen to think that if you want to see a bright future for humanity, space has got to be a part of the solution. Now, I don't believe that we need to leave the planet because we've done horrible things to it and the planet is no longer livable. That's not the point of view I come from at all. I simply think that we're seven billion people on this planet today, and if we think that humanity is a good thing, and we want to continue to expand and have high quality of life for everyone, we've got to get out there. Because everything that we consider as being precious on Earth is plentiful out there in space. Any resources, anything that we've fought over on this planet, is plentiful in space. Energy, precious metals, even real estate—it's all out there. Why wouldn't we go out there if we want to see a larger expansion of humanity? So yeah, I'm a strong believer that the two go hand in glove.

**A. NEWITZ** Lauren, what do you think? What's going to happen when we get out into space? Are going to keep making the same mistakes? Are we going to be able to get past our problems?

**L. OLIVER** Well, first of all, I'm a space nerd, but there are no dogs in space, so I'm skeptical.

**A. NEWITZ** Are you sure?

**L. OLIVER** There are no dogs in space! I've described myself this week as a "There is no Planet B" knuckle dragger, because I think that space exploration should be knowledge. It should be driven by knowledge, not driven by the idea that a few billionaires will be able to go to Mars while the rest of the planet fries. I'm starting to see a disconnect in the discussion between space travel and why we do it. We do it to uplift all of us. It's not an escape hatch. Let's face it: seven billion people are not going to fly away in escape pods and eat space sticks for the rest of their lives. Also, there are no dogs in space.

**A. NEWITZ** And there are no dogs in space.

**L. OLIVER** There are no dogs.

.................................................

## There are no dogs in space.

.................................................

**A. NEWITZ** I thought it was really interesting yesterday that Neal Stephenson brought up the show *The Expanse*, which is about what happens in the Solar System when we've colonized the inner planets and the asteroid belt. There are some sort of satellite colonies in the outer planet as well. But one of the things that happens in that show is that we start mining asteroids, which sounds great; there's been a lot of panels here where people have said, "Yes! We're going to get resources from asteroids." But what happens in the show is all the Belters who are living out there mining the asteroids basically become the future coal miners of Virginia in the early twentieth century and late nineteenth century. They're horribly abused. They live in substandard conditions. They don't have enough money to pay for gravity, so they have all kinds of bone problems. They're impoverished. They have terrible food. So we get all the resources, but then we still have this same problem that we have on Earth between the haves and the have-nots.

**L. OLIVER** Colonialism.

**A. NEWITZ** We still have colonialism, but it's in space, so it's all better. How do we address those problems? How do we think we can go into space in a way that doesn't replicate the problems that we've had on Earth?

**A. ELLIS** Well, I mean, first of all, we're going to be making mistakes.

**A. NEWITZ** Yes.

**A. ELLIS** Okay. Let's just acknowledge that progress at some point is going to have ups and downs.

**L. OLIVER** Two steps forward, two steps back.

**A. ELLIS** Exactly. It's inevitable. It's the way it's always been. I think we as a society should be more comfortable with the idea that everything we do that's worth doing implies risk. Sometimes that risk is going to be catastrophic, and sometimes that risk is going to pay off. The reason why we are where we are and why we're hopefully going to be heading in a positive direction is because we're willing to take those risks and be optimistic about it. Also, we've got to understand that the problems we're going to have in the future are fundamentally different than the types of problems that we have today. I bet you that there are going to be all sorts of problems that we're going to have in the future we haven't even thought about. As brilliant as your ideas are, crazy things are going to happen.

**A. NEWITZ** I'm okay with new problems. That sounds good.

**A. ELLIS** Yeah. That's kind of where I'm leaning.

**L. OLIVER** And, hopefully, not new problems that destroy the Earth where we're like, If only we'd known . . .

**A. ELLIS** Hopefully not.

**L. OLIVER** Yeah.

**A. NEWITZ** Lauren, do you think that we really need to first solve our environmental problems on Earth before we can start going into space? Is that your trajectory?

**L. OLIVER** I'm a knuckle dragger, but I'm not that bad. No. No, the problem we have is a problem of allocation. It's not that we can't grow food for everybody or that we can't educate everybody. I don't see it as "If we go to space, then we can't feed people." That's just a matter of how our political system starves some people and rewards others. We shouldn't be fooled into thinking that we have to make these kinds of choices, especially with the climate change issue.  -273-

They say, "Well, we can either have the environment or we can have jobs." We absolutely know that to be completely false! When we look at the difference in how many people are employed by the coal industry in this country and the amount of people that are employed by the expanding solar industry—I could give you the numbers except that I don't have them, but it's like this much versus this much. I just don't buy any of those arguments. I think we're being told things that are just not true.

**A. ELLIS** Can I just add to that?

**L. OLIVER** Yeah, yeah, please.

**A. ELLIS** I'm curious, who here is a *Star Trek* fan?

APPLAUSE. WHOOPS. HOLLERS.

**A. ELLIS** Okay. Are you all familiar with the replicator?

APPLAUSE

**A. ELLIS** Yeah? We were having this discussion earlier today, and we thought one thing that could make a substantial difference in the way that we live, a massive shock to the economy as it's structured today, would be the creation of something like the replicator. In the same way that the internet democratized access to information, maybe material goods would undergo the same kind of disruption.

**L. OLIVER** Well, we can transition into 3D printers, right? You can just download what you need from an open site. What if we put our recycled plastic into that machine, because we're failing at recycling plastic? But what if it went into the machine? What we talked about earlier just made me crazy because you said, "Oh, no. It would be at an atomic level, where we could just pull out atoms, converge them, drop it in, and then you have a cheesecake." Just like in *Star Trek*. GJJZZZZZZZzzzzzz. Cheesecake. Okay.

**A. NEWITZ** I mean, that's the premise of Cory Doctorow's novel *Walkaway*—

**L. OLIVER** Really?

**A. NEWITZ** —that 3D printers become the precursor technology for having a more democratized access to material goods, but also to political choice. That somehow having access to those material goods is what we need to finally have a fully realized democracy where everyone is able to participate to the best of their ability and everyone feels like they have a stake in the culture that they're building. I was going to say, I'm a big fan of Mae Jemison's 100 Year Starship project, which you can check out online. She's a former astronaut and she's very interested in this idea of long-term projects to get off the Earth that start by solving environmental problems on Earth along the way. The idea is, if we can solve questions about, for example, remediating damage to an ecosystem, that could lead into creating one of these autonomous ecosystems that we heard about yesterday that could become part of a long-term starship voyage.

**A. NEWITZ** We're closing in on the end, so I wanted to ask you guys a final question, which is: What gives you hope for the future of humanity? And it can be a long answer or a short answer.

**A. ELLIS** Wow. You know, I feel like humanity's good. I think that we ... If you think about how old civilization actually is, we're not that old; we're just making stupid mistakes.

**A. NEWITZ** We're just starting.

**A. ELLIS** We are, and we're making stupid mistakes. But that's all right. We should be a little more forgiving to ourselves. Let's just take that into consideration and understand that the future is actually quite bright. As long as we change a couple of things and mature along the way so that we don't self-destruct, I think humanity is generally heading in the right direction. I've got a couple of ideas for how we can go about doing this. One of them would be maybe to be less offended when we encounter ideas that are a little bit strange to us or things that we initially might not agree with. So that might be one good thing.

-275-

Another thing is that we could be a bit more curious, especially about different points of view. Where are they coming from? Can we be more empathetic? Can we be more kind? Can we be more loving? Another thing is to maybe not take ourselves so seriously. I know this panel is about the end of the world and everything, but maybe it helps to have a little bit of a sense of humor and context about where we are and not be so hard on ourselves.

## What gives you hope for the future of humanity?

**A. NEWITZ** We could be a little bit more humble. Maybe we're not the center of everything.

**L. OLIVER** I know we're not the center of everything.

**A. NEWITZ** What's your hope?

**L. OLIVER** I'm an art idiot, and so I have sought out scientists and activists in this area, especially ocean people and ocean polar people. My total rockstar hero is a woman named Dr. Sylvia Earle.

APPLAUSE

**L. OLIVER** Yes. Please applaud this woman and see her movie on Netflix, *Mission Blue*. You will come away an ocean defender. She's amazing. I think she just turned eighty-nine and she goes around the world all the time and she tries to get people into the ocean, just get them to the water so that they can understand what we're trying to save. The whole point is to save what you love, and if you bring people in contact and let them enjoy the world—just please enjoy the world—then you'll want to save it. She's great. She creates "Hope Spots," marine sanctuaries. There's a fellow in England, maybe you know him, George Monbiot. He's starting to develop ideas about rewilding areas, because, really, that's what's sustainable. Real ecologies sustain our existence, not that.

In general, my art project, what I'm trying to wrestle with now, is how do we evolve from what we are to what we want to be? What steps do we take, and how do we help each other do it?

**A. NEWITZ** I just want to end by saying, since you were mentioning evolution, that one of the profound things that we learn from evolution is that life-forms want to survive. The thing that I find so hopeful about humans is that when they read about the history of disasters—whether it's a civilization-level disaster or a train wreck, or an earthquake—when humans have the urge to survive, they always want to take people with them. People will suddenly save someone from a burning truck, or they'll build a fruit stand in the park and give away free fruit after an earthquake. This is built into us. We want to survive with friends. I think that is a profoundly hopeful part of how we're built as biological creatures. It extends outward to our ecosystems. We're not going to get to survive with friends if we don't bring our bubble of life with us into the future along with all of the things that we eat that grow on the planet and all the animals that eat the things that grow on the planet. You know?

I think just that urge to survive, and to survive along with friends, is what really gives me hope. So with that, thanks very much for being at this amazing conference!

**L. OLIVER** Thank you.

**A. NEWITZ** Have hope for the future! Think of ways that you can intervene and maintain hope, because that's the only way we're going to fix things. It's how we're going to get out of our paralysis.

**L. OLIVER** Thank you.

DAVID C. KRAKAUER is the President and William H. Miller Professor of Complex Systems at the Santa Fe Institute. He works on the evolution of intelligence and stupidity on Earth, where the first is admired but rare; the second is feared but common.

CAITLIN L. McSHEA is the InterPlanetary Festival Director and a Miller Omega Program manager at the Santa Fe Institute. She double-majored in biology and philosophy, and has a master's degree in liberal arts from St. John's College. She dabbles in botanical illustration.

## THE SANTA FE INSTITUTE PRESS

The SFI Press endeavors to communicate the best of complexity science and to capture a sense of the diversity, range, breadth, excitement, and ambition of research at the Santa Fe Institute. To provide a distillation of discussions, debates, and meetings across a range of influential and nascent topics.

*To change the way we think.*

### SEMINAR SERIES
New findings emerging from the Institute's ongoing working groups and research projects, for an audience of interdisciplinary scholars and practitioners.

### ARCHIVE SERIES
Fresh editions of classic texts from the complexity canon, spanning the Institute's thirty years advancing the field.

### COMPASS SERIES
Provoking, exploratory volumes aiming to build complexity literacy in the humanities, industry, and the curious public.

### ALSO FROM SFI PRESS
*History, Big History, & Metahistory*
David C. Krakauer, John Lewis Gaddis, & Kenneth Pomeranz, eds.

*The Emergence of Premodern States*
*New Perspectives on the Development of Complex Societies*
Jeremy A. Sabloff & Paula L. W. Sabloff, eds.

*Emerging Syntheses in Science*
*Proceedings of the Founding Workshops of the Santa Fe Institute*
David Pines, ed.

*For forthcoming titles, inquiries, or news about the Press, contact us at*
SFIPRESS@SANTAFE.EDU.

## ABOUT THE SANTA FE INSTITUTE

The Santa Fe Institute is the world headquarters for complexity science, operated as an independent, nonprofit research and education center located in Santa Fe, New Mexico. Our researchers endeavor to understand and unify the underlying, shared patterns in complex physical, biological, social, cultural, technological, and even possible astrobiological worlds. Our global research network of scholars spans borders, departments, and disciplines, bringing together curious minds steeped in rigorous logical, mathematical, and computational reasoning. As we reveal the unseen mechanisms and processes that shape these evolving worlds, we seek to use this understanding to promote the well-being of humankind and of life on Earth.

## COLOPHON

The body copy for this book was set in EB Garamond, a typeface designed by Georg Duffner after the Ebenolff-Berner type specimen of 1592. Headings are in Kurier, a typeface created by Janusz M. Nowacki, based on typefaces by the Polish typographer Małgorzata Budyta. Additional type is set in Cochin, a typeface based on the engravings of Nicolas Cochin, for whom the typeface is named.

The SFI Press complexity glyphs used throughout this book were designed by Brian Crandall Williams.

ZERO

ONE

TWO

THREE

FOUR

FIVE

SIX

SEVEN

EIGHT

NINE

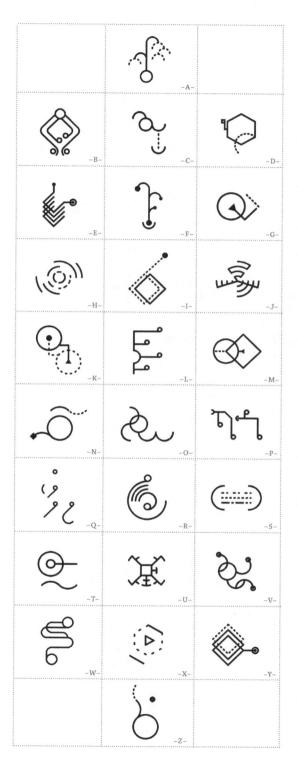

-A-

-B-

-C-

-D-

-E-

-F-

-G-

-H-

-I-

-J-

-K-

-L-

-M-

-N-

-O-

-P-

-Q-

-R-

-S-

-T-

-U-

-V-

-W-

-X-

-Y-

-Z-

COMPASS SERIES

CPSIA information can be obtained
at www.ICGtesting.com
Printed in the USA
LVHW090933180120
643426LV00003BA/69/J